# Power System Load Frequency Control
## Classical and Adaptive Fuzzy Approaches

T0314422

# Power System Load Frequency Control
## Classical and Adaptive Fuzzy Approaches

Hassan A. Yousef

Department of Electrical and Computer Engineering,
Sultan Qaboos University,
Muscat, Oman

CRC Press
Taylor & Francis Group
Boca Raton London New York

CRC Press is an imprint of the
Taylor & Francis Group, an **informa** business

CRC Press
Taylor & Francis Group
6000 Broken Sound Parkway NW, Suite 300
Boca Raton, FL 33487-2742

First issued in paperback 2020

ISBN 13: 978-0-367-57389-8 (pbk)
ISBN 13: 978-1-4987-4557-4 (hbk)

**Library of Congress Cataloging-in-Publication Data**

Names: Yousef, Hassan A., author.
Title: Power system load frequency control / Hassan A. Yousef.
Description: Boca Raton : Taylor & Francis, a CRC title, part of the Taylor & Francis imprint, a member of the Taylor & Francis Group, the academic division of T&F Informa, plc, [2017] | Includes bibliographical references and index.
Identifiers: LCCN 2016037377 | ISBN 9781498745574 (hardback : alk. paper) | ISBN 9781498745581 (e-book : alk. paper)
Subjects: LCSH: Electric power systems--Control. | Electric power-plants—Load
Classification: LCC TK1007 .Y68 2017 | DDC 621.31/7--dc23
LC record available at https://lccn.loc.gov/2016037377

**Visit the Taylor & Francis Web site at
http://www.taylorandfrancis.com**

**and the CRC Press Web site at
http://www.crcpress.com**

*To my late father, Abdel Halim Hassan Yousef;*

*To my mother, Kawthar Al Gebaly;*

*To my grandchildren, Yomna and Malik*

# Contents

# *Preface*

Several years ago, I started thinking about writing this textbook. Actually, the textbook project began when I joined Sultan Qaboos University (SQU), Oman, in 2009. This textbook is an outcome of separate lecture notes for classes I taught at the Department of Electrical Engineering, Alexandria University, Egypt, and the Department of Electrical and Computer Engineering, SQU, Oman. My goal was to write something thorough and readable. This textbook is intended for use in senior undergraduate or first-year postgraduate courses in power system control. Besides the academic readership, the book is also intended to benefit practitioners engineers and researchers looking for elaborated information on the active power frequency control of power systems.

As implied by the book's title, load frequency control of power systems is the main theme. The book presents a balanced blend of classical and intelligent load frequency control techniques. These techniques are presented in a systematic way to help the readers thoroughly understand the different control concepts. The classical control techniques introduced in this book include PID, pole placement, observer-based state feedback, and static and dynamic output feedback controllers. On the other hand, the intelligent control techniques explained here are of adaptive fuzzy control type. The only assumed prerequisites for smooth reading and grasping the ideas presented in the book are undergraduate fundamental courses in linear control systems and power system analysis.

The book consists of ten chapters organized in two parts and one appendix. Part I presents modeling and classical control approaches for load frequency control problem. This part is covered in Chapters 1 through 5. Chapters 6 through 10 constitute Part II that presents adaptive fuzzy logic control approaches. Chapter 1 introduces the decoupling property of power system control loops, namely, the active power frequency (Pf) and reactive power voltage (QV) channels. Also, the modeling of different components and the primary and secondary control loops associated with load frequency control (LFC) scheme are presented in this chapter. Chapter 2 presents state-variable models of multi-area power systems. These models are used to design different classical LFCs such as pole placement, observer-based state feedback, and dynamic output feedback. The deregulation of power systems that started in the early 1990s is explained in Chapter 3. In this chapter, different participation factors, such as distribution and generation participation factors, are explained and the block diagram model of LFC schemes for a multi-area deregulated power system is constructed. Chapter 4 provides classical control design using the transfer function approach. Different architectures of PID controllers along with the well-known Ziegler–Nichols tuning methods are highlighted. An analytical design of PID is presented based on internal model control (IMC). Furthermore, this chapter presents model reduction using the half-rule method and PID controller design based on the reduced models. Chapter 5 describes design of decentralized LFC for multi-area power systems. An independent controller for an isolated area is designed using first-order plus time delay (FOPTD) and second-order plus time delay (SOPTD) approximate models of an open-loop plant. Then, the stability of a closed-loop interconnected system is investigated using structured singular value (SSV). Decentralized controller design in the presence of communication time delay is also given in this chapter.

Adaptive fuzzy load frequency control is covered in Part II. Chapter 6 is an introductory chapter where the different types of fuzzy logic systems used in control applications and function approximation are presented. Fuzzy basis function (FBF) expansion of the most widely used fuzzy logic system is highlighted, and its property in capturing the local and global characteristics of functions to be approximated is explained. Chapter 7 presents two nonadaptive fuzzy control techniques: stable fuzzy control and optimal fuzzy control. The conditions for global exponential stability of a closed-loop fuzzy system are given for stable control technique. The classical linear quadratic regulator (LQR) problem is formulated and solved using the principle of optimality and its Hamilton–Jacobi–Bellman (HJB) equation result. The optimal fuzzy control is then solved using the algebraic Riccati equation (ARE). The LFC of an isolated power system is designed using the presented stable and optimal fuzzy techniques. Chapter 8 describes two different categories of adaptive fuzzy logic control techniques: direct adaptive fuzzy control (DAFLC) and the indirect adaptive fuzzy logic control (IAFLC). Both techniques involve a fuzzy logic system and a mechanism for parameter adaptation. In IAFLC, the fuzzy logic system is used to model an unknown plant; while in DAFLC, the fuzzy logic system is employed directly as a controller.

The design of LFC schemes for a multi-area power system based on the DAFLC and IAFLC techniques are developed in Chapters 9 and 10, respectively. The design is based on the controller canonical form (CCF) of the state model for a multi-area system. The transformation to the CCF is illustrated using similarity transformation. Lyapunov function synthesis approach is used to derive the parameter adaptation rules as well as to prove the boundedness of a tracking error. Also, a method to attenuate the tracking error for two types of controllers in an $H_\infty$ sense is introduced. A two-area power system is given in Chapters 9 and 10 as an illustrative example. A system equipped with a DAFLC and an IAFLC is simulated under the effect of step load disturbance. Finally, the Appendix presents MATLAB® codes for the examples presented in Chapter 2.

I have tried my best to make this book error free. However, some typographical errors may remain. Please e-mail your comments to hyousef@ squ.edu.om. I would very much appreciate to hear of any errors found by the readers.

**Hassan A. Yousef**
*Muscat, Oman*

MATLAB® is a trademark of The MathWorks, Inc. and is used with permission. The MathWorks does not warrant the accuracy of the text or exercises in this book. This book's use or discussion of MATLAB® software or related products does not constitute endorsement or sponsorship by The MathWorks of a particular pedagogical approach or particular use of the MATLAB® software.

# Acknowledgments

It is my duty to thank many people who have helped, in one way or another, to make this book a reality. I thank the publisher representatives for their constant encouragement and support. I am grateful to Professor M. Simaan, of the University of Central Florida, USA, who taught me control theory and how to do research. I am also indebted to Professor O. Sebakhy of Alexandria University, Egypt, who introduced me to the field of control engineering.

I had fruitful discussions with and received useful feedback from many postgraduate students, including Dr. M. Al-Khatib of Sandia National Labs, USA, Dr. M. Hamdy of Menofia University, Egypt, Khalfan Al-Kharusi of DCRP, Oman, and Saif Al-Kalbani of PDO, Oman.

The support from Sultan Qaboos University (SQU) is highly appreciated. Special thanks go to Professor Abdullah Al-Badi, Dean of Engineering, SQU, for his support and encouragement. I also thank my colleagues from SQU, Dr. M. Shafiq, Dr. M. Al Badi, Dr. M. H. Soliman, Dr. A. Al-Hinai, and Dr. R. Al Abri.

Finally, I thank my family Hanan, Amr, Selma, Eslam, Osama, and Hosam for encouraging me to write this book. Their love, patience, and continued support were the main reasons for its successful completion.

# *Author*

**Hassan A. Yousef** received his BSc (honor) and MSc in electrical engineering from Alexandria University, Alexandria, Egypt, in 1979 and 1983, respectively. He earned his PhD in electrical and computer engineering from the University of Pittsburgh, Pennsylvania, in 1989. He spent 15 years in Alexandria University as assistant professor, associate professor, and professor. He worked at Qatar University for six years as assistant professor and then associate professor. In Qatar University, he held the position of acting head of the Electrical and Computer Engineering Department. He was a visiting associate professor at the University of Florida, Gainesville, in summer 1995. Now he is with the Department of Electrical and Computer Engineering, Sultan Qaboos University, Sultanate of Oman, Muscat. He supervised 28 completed MSc theses and 8 completed PhD dissertations. Dr. Yousef published 100 papers in refereed journals and international conferences in the area of control system, nonlinear control, adaptive fuzzy control, and intelligent control applications to power systems and electric drives.

# Part I

# Modeling and Classical Control Methods

# 1

## Load Frequency Control of Power Systems

### 1.1 Control Problems of Power Systems

A power system consists of classical generators or renewable energy sources, loads, and transmission and distribution networks. The installed generating capacity of a power system must equal the peak load demand plus a spinning reserve. The spinning reserve comprises at least one lightly loaded generator in the system. A power system is connected through tie-lines to import/export power to/from neighboring systems.

The key equations to analyze a power system are the active and reactive power flow equations. Consider the $i$th bus of an $N$-bus power system. The bus voltage has a magnitude $|V_i|$ and phase angle $\delta_i$, that is,

$$V_i = |V_i| \angle \delta_i \tag{1.1}$$

and the bus current $I_i$ is given as

$$I_i = I_{Gi} - I_{Li} = |I_i| \angle \gamma_i \tag{1.2}$$

where $I_{Gi}$ and $I_{Li}$ are the total generation current and the total load current connected to the $i$th bus, respectively. The active and reactive powers $(P_i, Q_i)$ at this specific bus are determined as

$$P_i = |V_i||I_i| \cos(\delta_i - \gamma_i) \tag{1.3}$$

$$Q_i = |V_i||I_i| \sin(\delta_i - \gamma_i) \tag{1.4}$$

Using the bus admittance matrix $Y_{Bus} = \begin{bmatrix} y_{11} & \cdots & y_{1n} \\ \vdots & \ddots & \vdots \\ y_{n1} & \cdots & y_{nn} \end{bmatrix}$ and the bus volt-

age $V_{Bus} = \begin{bmatrix} V_1 \\ \vdots \\ V_N \end{bmatrix}$, the bus current $I_{Bus}$ can be written as

$$I_{Bus} = Y_{Bus}V_{Bus} = \begin{bmatrix} I_1 \\ \vdots \\ I_N \end{bmatrix} \tag{1.5}$$

From (1.5), the $i$th bus current in terms of the elements of the bus admittance matrix $y_{ij} = |y_{ij}| \angle \rho_{ij}$ is given by

$$I_i = \sum_{k=1}^{N} |y_{ik}||V_k| \angle (\rho_{ik} + \delta_k) \tag{1.6}$$

The complex power $S_i$ at the bus $i$th bus takes the form

$$S_i = P_i + jQ_i = V_i I_i^* \tag{1.7}$$

where $I_i^*$ is the conjugate of the current $I_i$. Upon substitution of (1.6) in (1.7), we obtain the following expressions of the powers:

$$P_i = \sum_{k=1}^{N} |y_{ik}||V_k||V_i| \cos (\delta_i - \rho_{ik} - \delta_k) \tag{1.8}$$

$$Q_i = \sum_{k=1}^{N} |y_{ik}||V_k||V_i| \sin (\delta_i - \rho_{ik} - \delta_k) \tag{1.9}$$

A well-designed power system will operate in a normal state with the following requirements [1]:

1. The load demand of active and reactive power should be met and the power flow equations (1.8) and (1.9) are satisfied.
2. The frequency and bus voltage magnitude are to be kept within specified standard limits.
3. Different components in the system are not to be overloaded.

The control engineers are striving to maintain the power system in a normal state. This task is not straightforward due to the complexity of power system structure. However, the noninteraction property of a power system makes the control task somewhat simple. The noninteraction property is presented in the next section.

## 1.2 Decoupling Property of Power System Control

Consider small changes in the active and reactive powers $\Delta P_i$, $\Delta Q_i$. These changes will result in variations in the voltage magnitude $\Delta \lvert V_i \rvert$ and in the angle $\Delta \delta_i$. These variations can be determined using perturbation analysis. To this effect, we rewrite (1.8) and (1.9) as

$$h_i = P_i - \sum_{k=1}^{N} \lvert y_{ik} \rvert \lvert V_k \rvert \lvert V_i \rvert \cos \left( \delta_i - \rho_{ik} - \delta_k \right) \tag{1.10}$$

$$g_i = Q_i - \sum_{k=1}^{N} \lvert y_{ik} \rvert \lvert V_k \rvert \lvert V_i \rvert \sin \left( \delta_i - \rho_{ik} - \delta_k \right) \tag{1.11}$$

At a given operating point $(P_{io}, Q_{io}, \lvert V_{io} \rvert, \delta_{io})$ these two equations become zero. For steady-state perturbations $\Delta P_i$, $\Delta Q_i$, $\Delta \lvert V_i \rvert$, and $\Delta \delta_i$ around the operating point, the perturbations in $h_i$ and $g_i$ can be written as

$$\Delta h_i = \frac{\partial h_i}{\partial P_i} \Delta P_i + \sum_{i=1}^{N} \frac{\partial h_i}{\partial \delta_i} \Delta \delta_i + \sum_{i=1}^{N} \frac{\partial h_i}{\partial \lvert V_i \rvert} \Delta \lvert V_i \rvert = 0 \tag{1.12}$$

$$\Delta g_i = \frac{\partial g_i}{\partial P_i} \Delta Q_i + \sum_{i=1}^{N} \frac{\partial g_i}{\partial \delta_i} \Delta \delta_i + \sum_{i=1}^{N} \frac{\partial g_i}{\partial \lvert V_i \rvert} \Delta \lvert V_i \rvert = 0 \tag{1.13}$$

In a typical power system, the bus admittance elements are highly inductive ($\rho_{ik} = 90°$), and the line power angle $(\delta_i - \delta_k) \leq 30°$. Under these situations, it can be easily shown that the bus voltage angle variation $\Delta \delta_i$ is affected mainly by the active power change $\Delta P_i$, while the magnitude of the bus voltage can be controlled by the variation in the reactive power change $\Delta Q_i$. This means that the active and reactive power can control the bus angle and the voltage magnitude independently. In fact, there is a minor impact of the active power control on the bus voltage magnitude and likewise the reactive power control on the bus voltage angle. This fact is illustrated in the following example.

### Example 1.1

Consider the two-bus power system shown in Figure 1.1.
    Express the changes in the bus voltage magnitudes $\Delta \lvert V_1 \rvert$ and $\Delta \lvert V_2 \rvert$ and the changes in the phase angles $\Delta \delta_1$ and $\Delta \delta_2$ in terms of the steady-state changes in bus active and reactive power, if $V_1 = 1.05 \angle 5°$ pu, $V_2 = 0.98 \angle 10°$ pu, and the transmission line is lossless with a reactance of $X_L = 0.1$ pu.

**FIGURE 1.1**
Two-bus power system.

**Solution:**

For this example:

$$h_1 = P_1 - \frac{|V_1||V_2|}{X_L}\sin(\delta_1 - \delta_2), \qquad h_2 = P_2 - \frac{|V_1||V_2|}{X_L}\sin(\delta_1 - \delta_2),$$

$$g_1 = Q_1 - \frac{|V_1||V_2|}{X_L}\cos(\delta_1 - \delta_2) \text{ and } g_2 = Q_2 - \frac{|V_1||V_2|}{X_L}\cos(\delta_2 - \delta_1).$$

Upon evaluation of (1.12) and (1.13), we get

$$\begin{bmatrix} 1 & 0 & 0 & 0 \\ 0 & 1 & 0 & 0 \\ 0 & 0 & 1 & 0 \\ 0 & 0 & 0 & 1 \end{bmatrix}\begin{bmatrix} \Delta P_1 \\ \Delta P_2 \\ \Delta Q_1 \\ \Delta Q_2 \end{bmatrix} + \begin{bmatrix} -10.25 & -10.25 & 0.85 & 0.91 \\ 10.25 & -10.25 & -0.85 & -0.91 \\ -0.9 & 0.9 & -9.76 & -10.46 \\ -0.9 & 0.9 & 9.76 & -10.46 \end{bmatrix}\begin{bmatrix} \Delta\delta_1 \\ \Delta\delta_2 \\ \Delta|V_1| \\ \Delta|V_2| \end{bmatrix} = 0.0$$

which gives

$$\begin{bmatrix} \Delta\delta_1 \\ \Delta\delta_2 \\ \Delta|V_1| \\ \Delta|V_2| \end{bmatrix} = \begin{bmatrix} -0.049 & 0.049 & -0.0084 & 0 \\ -0.049 & -0.049 & 0 & 0 \\ 0 & 0 & -0.051 & 0.051 \\ 0 & -0.0083 & -0.047 & -0.047 \end{bmatrix}\begin{bmatrix} \Delta P_1 \\ \Delta P_2 \\ \Delta Q_1 \\ \Delta Q_2 \end{bmatrix}$$

These results demonstrate the decoupling effects of the active power and reactive power control. In other words, the equations $\Delta\delta_1 \cong -0.049\Delta P_1 + 0.049\Delta P_2$, $\Delta\delta_2 = -0.049\Delta P_1 - 0.049\Delta P_2$, $\Delta|V_1| = -0.051\Delta Q_1 + 0.051\Delta Q_2$, and $\Delta|V_2| \cong -0.047\Delta Q_1 - 0.047\Delta Q_2$ mean that the incremental change in active power affects mainly the incremental change in phase angle and the incremental change in reactive power affects the change in voltage magnitude.

The following observations are in effect [2]:

1. Steady-state changes in the real power $\Delta P_i$ mainly affect the bus voltage phase angles and the active power line flows.
2. Steady-state changes in the reactive power $\Delta Q_i$ affect only the bus voltage magnitudes and thus the reactive power line flows.

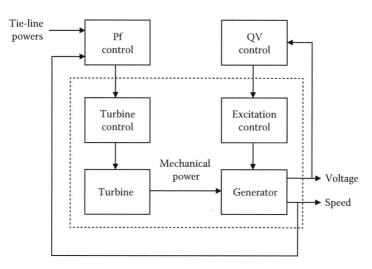

**FIGURE 1.2**
Control channels of a generating unit.

Based on these analyses and observations, we conclude that the power system control tasks can be divided into two separate control channels. These channels are depicted in Figure 1.2 as Pf and QV controls. The Pf (power frequency) control channel is responsible for adjusting the turbine shaft power in order to regulate the speed (and hence the frequency) and control the active power output. On the other hand, the function of the QV control channel is to regulate the generator voltage and control the reactive power output. In this book, we concentrate only on the modeling and control of the Pf control loop.

## 1.3 Power-Frequency Control Channel

Ensuring constant frequency is a crucial factor for satisfactory operation of power systems and is essential for industry operation and production. The frequency should be constant or experience small perturbations around the operating value as prescribed by the electric power utility. The frequency and active power balance in a system are dependent as explained in Section 1.2. A change in the active load power at any given bus of a system will produce an imbalance between the generating power and the load power. The stored kinetic energy of the generator supplies this power difference. As a result, the generator speed falls and hence the frequency. Because there are many generating units in a power system, some means must be provided to allocate the load changes to the generators.

The change in frequency is sensed and fed back through a speed governor to control the inlet valve position of the turbine. This in turn changes the input to the turbine and hence the output of the generator in order to reduce the active power mismatch. This speed governor is termed as the primary control. The primary control action provides a relatively fast but coarse frequency control. The response time of this loop is limited by the inherent turbine time constant. In an interconnected power system having more than one area, the generation within each area has to be controlled to maintain scheduled power interchange. The control of generation and frequency is commonly termed as load-frequency control (LFC). The LFC has a secondary control action working in a slow reset mode to eliminate the frequency error and control the power interchange between generators.

## 1.3.1 Primary Control Action

The governors are devices attached to the turbines to change the mechanical input power to the generators. The input to the governor is the speed deviation $\Delta\omega$ (due to load change) defined as the difference between the generator speed $\omega_a$ and the reference speed $\omega_r$, that is, $\Delta\omega=\omega_a-\omega_r$. This signal is amplified and integrated to produce an actuating valve/gate signal $\Delta P_g$. The actuating signal $\Delta P_g$ is responsible for providing a change in the inlet steam for steam turbines or water flow rate for hydro turbines. The governor action can be modeled as shown in Figure 1.3.

For stable load division between generating units existing in a power system, the governor of each generator will have a regulation characteristic such that the speed falls as the load increases. This speed droop relation can be obtained by adding a feedback around the integrator as shown by the dotted lines in Figure 1.3. The transfer function model of a governor equipped with a speed droop characteristic can be written as

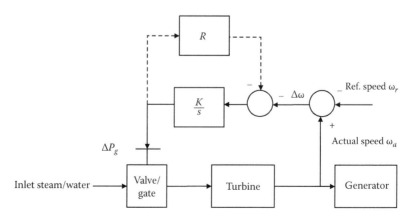

**FIGURE 1.3**
The governor action.

$$\frac{\Delta P_g}{\Delta \omega} = \frac{-(1/R)}{1+sT_g} \qquad (1.14)$$

where
$T_g = 1/KR$ is the governor time constant
$R$ is the speed droop regulation constant

At steady state, it is obvious from (1.14) that

$$\Delta \omega = -R\Delta P_g \qquad (1.15)$$

This means that the variation in valve/gate opening ($\Delta P_g$), accordingly the generator output power ($\Delta P$), is proportional to speed (frequency) variation ($\Delta \omega$). Therefore, the generator frequency and power are related according to

$$f = -RP \qquad (1.16)$$

This relation is depicted in Figure 1.4, where $f_0$ and $f_{fl}$ are the no-load and full-load frequencies, respectively, and $P_0$ is the full-load power. The unit of $R$ is taken as Hz per pu MW.

If two or more generators having a speed droop regulation are connected to a power system, there will be a unique frequency at which they will share a load change. For example, consider two generators with speed droop regulation constants $R_1$ and $R_2$. The ratio between the amount of load shared by each generator due to a load increase $\Delta P_L$ is given by

$$\frac{\Delta P_1}{\Delta P_2} = \frac{R_2}{R_1} \qquad (1.17)$$

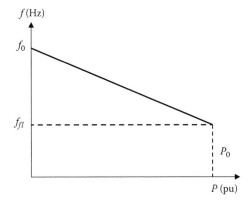

**FIGURE 1.4**
Governor frequency droop characteristic.

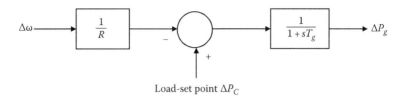

**FIGURE 1.5**
Block diagram representation of a typical governor.

In practice, the relationship between speed and load can be adjusted by changing a load set point through a speed-changer motor. A block diagram representation of a typical governor is given in Figure 1.5.

### 1.3.2 Turbine Model

The turbine is responsible for converting the valve/gate position into mechanical power on the shaft coupled to the generating unit. There are many types of turbines that can exist in a power system such as non-reheat steam, reheat steam, hydro, gas, and combined cycle turbines. The transfer function model of each of these turbines will be discussed in this section.

### 1.3.2.1 Steam Turbine Model

A steam turbine converts stored energy of high pressure and high temperature steam into mechanical energy. A steam turbine may be of either reheat or non-reheat type. In a reheat-type turbine, when the steam leaves the high-pressure section of the turbine, it returns back to the boiler. In the boiler, it gets reheated before entering the intermediate-pressure section of the turbine. This reheating action improves turbine efficiency. A simplified transfer function for a reheat-type steam turbine can be written as [3].

$$\frac{\Delta P_m}{\Delta P_g} = \frac{1 + sKT_r}{(1 + sT_t)(1 + sT_r)} \tag{1.18}$$

where
$\Delta P_m$ is the change in turbine mechanical power output
$\Delta P_g$ is the change in steam valve position
$K$ is percentage of the total turbine power generated by the high-pressure section
$T_r$ is the time constant of the reheater
$T_t$ is the time constant of the inlet steam

For a non-reheat type, $T_r = 0$, and (1.18) reduces to

$$\frac{\Delta P_m}{\Delta P_g} = \frac{1}{1 + sT_t} \tag{1.19}$$

**Example 1.2**

Draw the block diagram of a governor and reheat turbine system and then find the steady state in turbine power change due to a frequency change of 0.2%. The governor droop regulation is 2.5 Hz/pu MW. If $K=0.3$, $T_r=7.0$ s, $T_t = 0.3$ s, and $T_g=0.25$ s, plot $\Delta P_m(t)$ when $\Delta\omega(t)$ is a step input with amplitude 0.2%. Assume the nominal frequency is 50 Hz.

**Solution:**

A block diagram of a governor–turbine system model is shown in Figure 1.6, where $\Delta P_G$, $\Delta P_C$, and $\Delta P_L$ are the changes in generator power, the speed changer setting, and the load power, respectively.

The transfer function $\Delta P_m/\Delta\omega$ can be written as

$$\frac{\Delta P_m}{\Delta\omega} = \frac{(1/R)(1+sKT_r)}{(1+sT_g)(1+sT_t)(1+sT_r)}$$

where $\Delta P_C$ and $\Delta P_L$ are assumed to be zero. Therefore,

$$\Delta P_m\big|_{steady\ state} = \left(\frac{1}{R}\right)\Delta\omega\big|_{steady\ state} = \left(\frac{1}{2.5}\right) \times 0.2\% \times 50 = 0.04 \text{ pu MW. The time}$$

response of $\Delta P_m(t)$ is evaluated using MATLAB® and shown in Figure 1.7.

## 1.3.2.2 Gas Turbine Model

A gas turbine mainly converts the chemical energy of the fuel into heat energy, which is then converted into mechanical energy. A simple gas turbine system consists of three main parts: a compressor, a combustor, and a gas turbine. A gas turbine (GAST) simplified model is shown in Figure 1.8. This model neglects the time constants associated with the fuel valve response and the load limit response.

## 1.3.2.3 Combined Cycle Turbine Model

A combined cycle power plant (CCPP) consists of at least one gas turbine (GAST), a steam turbine (ST), a heat-recovery steam generator (HRSG), and an electric generator [4]. The detailed models of multishaft and single-shaft CCPPs can be found in [4,5].

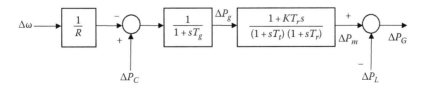

**FIGURE 1.6**
Governor and reheat steam turbine model.

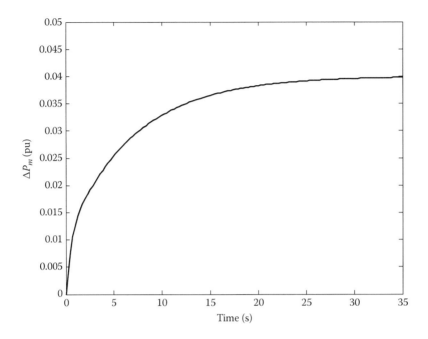

**FIGURE 1.7**
Transient response of turbine mechanical power due to step change in the speed.

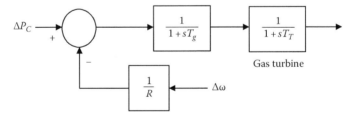

**FIGURE 1.8**
Block diagram of GAST basic standard model.

### 1.3.2.4 Hydro Turbine Model

The hydro turbine model can be derived using a small signal analysis of the nonlinear equations describing the water velocity $v$ through the penstock and the turbine mechanical power $P_m$. These equations are given by

$$v = K_v A\sqrt{h} \qquad (1.20)$$

$$P_m = K_m hv \qquad (1.21)$$

where

A is the gate position

h is the hydraulic head at the gate

$K_v$ and $K_m$ are constants

The transfer function of the linearized equations is given by [3]

$$\frac{\Delta P_m}{\Delta A} = \frac{1 - sT_w}{1 + 0.5sT_w} \tag{1.22}$$

where $T_w$ is the time required for a head $h_0$ to accelerate the water through the penstock from standstill to a velocity $v_0$.

As seen from (1.22), the model of the hydro turbine is a nonminimum phase (i.e., it has a zero on the right-hand side of the s-plane). Therefore, a governor with droop regulation constant R and time constant $T_g$, as given in Figure 1.5, will not be sufficient to ensure stable operation of the hydro turbine–generator set. In order to ensure stable operation with a satisfactory dynamic performance, a transient droop compensation should be used. One form of such compensation is given by

$$C(s) = \frac{1 + sT_a}{1 + sT_a\left(R_c/R\right)} \tag{1.23}$$

where $T_a$ and $R_c$ are chosen in a way to satisfy stability of the loop with a desired amount of damping. The optimum choice of these values is related to $T_w$ and the inertia constant H, as shown in the following equations:

$$R_c = \left(2.15 - 0.15T_w\right)\frac{T_w}{2H} \tag{1.24}$$

$$T_a = \left(4.5 - 0.5T_w\right)T_w \tag{1.25}$$

The block diagram of a hydro turbine along with the governor compensation is presented in Figure 1.9, where the generator and load are modeled as a first-order transfer function $\frac{1}{2Hs + D}$ (this will be shown in Section 1.3.3). The following example illustrates the benefit of using the governor compensator in association with a hydro turbine.

**Example 1.3**

Use the root locus technique to evaluate the effect of the speed governor with and without compensation for the hydro turbine connected to a generator. Use the following values of the system parameters: $T_w = 2.0$ s, $R = 0.05$, $T_g = 0.5$, $H = 5.0$, and $D = 1.0$.

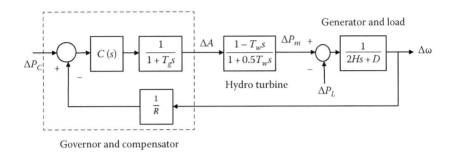

FIGURE 1.9
Hydro turbine with speed governor compensator.

**Solution:**

The loop transfer function of the system shown in Figure 1.9 without compensation can be written as

$$G(s) = \frac{K\left(\dfrac{1}{T_w} - s\right)}{\left(\dfrac{1}{T_g} + s\right)\left(\dfrac{2}{T_w} + s\right)\left(\dfrac{D}{2H} + s\right)} \tag{1.26}$$

where $K = 1/RT_g H$. The root locus command of MATLAB (*rlocus*) is used to plot the location of the closed-loop poles for $K \geq 0.0$. The result shown in Figure 1.10 indicates the following stability conditions: critically stable for $K = 1/RT_g H = 4.79$, stable for $K = 1/RT_g H \leq 4.79$, and unstable for $K = 1/RT_g H \geq 4.79$. This means that to ensure stable operation of the hydro turbine, the value of $R$ should satisfy $R \geq 0.0835$, which is larger than the typical value. To analyze the hydro-turbine system associated with the governor compensator using the root locus, we start by writing the closed-loop characteristic equation of the system shown in Figure 1.9. As a result, the characteristic equation is given by

$$1 + C(s)G(s) = 0 \tag{1.27}$$

The equivalent loop transfer function $G_c(s)$ in the presence of a governor compensator can be determined from (1.27) as

$$G_c(s) = \frac{R(1+sT_g)\left(1+s\dfrac{T_w}{2}\right)(D+2Hs)}{(1+sT_a)(1-sT_w)+sT_a R_c\left(1+sT_g\right)\left(1+s\dfrac{T_w}{2}\right)(D+2Hs)} \tag{1.28}$$

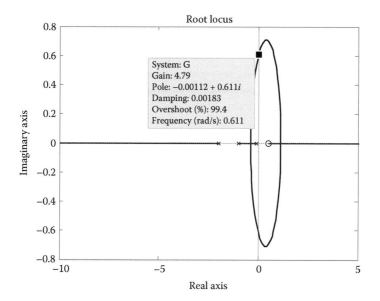

**FIGURE 1.10**
Root locus of a hydro turbine without a speed governor compensator.

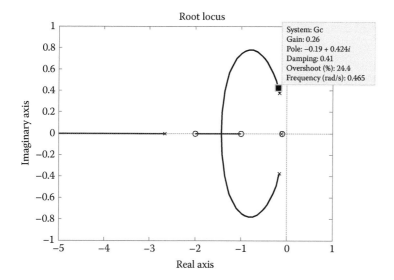

**FIGURE 1.11**
Root locus of a hydro turbine with a speed governor compensator.

where $T_a$ and $R_c$ are calculated from (1.24) and (1.25). Using the given values of the parameters, (1.28) is evaluated as

$$G_c(s) = \frac{0.26(s+0.1)(s+1)(s+2)}{(s+2.662)(s+0.1168)(s^2+0.3212s+0.1662)} \tag{1.29}$$

The root locus of (1.29) is shown in Figure 1.11. The closed-loop poles at the given values of the parameters are on the left-hand side of the s-plane indicating that the governor compensation enhances the turbine stability operation.

### 1.3.3 Generator and Load Model

The LFC problem is treated collectively by a harmony effort of all generator units within a control area. It is assumed that when a load variation exists in a control area, all generators belonging to this control area exhibit a coherent response. Hence, this group of generators can be represented as an equivalent generator. In what follows, we develop the generator dynamical model that can be used in the load frequency problem. We consider first the case of an isolated control area, and then the case of interconnected areas.

### 1.3.3.1 Generator Dynamic Model of an Isolated Control Area

Suppose the area experiences a load change of $\Delta \bar{P}_L$. Due to the turbine and governor actions, the mechanical power output of the generator increases by an amount $\Delta \bar{P}_m$. The incremental power balance $\left(\Delta \bar{P}_m - \Delta \bar{P}_L\right)$ will be acting as an accelerating power and dissipated in increasing the area kinetic energy $\bar{W}$ at a rate $d\bar{W}/dt$ and in increasing the load consumption. Since most of the power system loads are motors, the load consumption increase is assumed as $\bar{D}\Delta\bar{f}$, where $\bar{D} = \partial \bar{P}_L / \partial \bar{f}$ is the load damping constant expressed in MW/Hz and $\Delta\bar{f}$ is the frequency change in Hz. Thus, the power balance equation can be written as

$$\Delta \bar{P}_m - \Delta \bar{P}_L = \frac{d\bar{W}}{dt} + \bar{D}\Delta\bar{f} \tag{1.30}$$

The kinetic energy of the control area is proportional to the square of frequency (or speed), that is,

$$\bar{W} = W_0 \left(\frac{\bar{f}}{f_0}\right)^2 \tag{1.31}$$

where $W_0$ and $f_0$ are the nominal kinetic energy and nominal frequency, respectively. Assuming $\bar{f} = f_0 + \Delta\bar{f}$ and neglecting the second-order term, one can write (1.31) as

$$\bar{W} \cong W_0 \left( 1 + 2 \frac{\Delta \bar{f}}{f_0} \right) \tag{1.32}$$

Therefore

$$\frac{d\bar{W}}{dt} = 2 \frac{W_0}{f_0} \frac{d\Delta \bar{f}}{dt} \tag{1.33}$$

Dividing both sides of (1.33) by the base MVA (BMAV), we get

$$\frac{dW}{dt} = 2 \frac{W_0}{BMVA} \frac{d}{dt} \left( \frac{\Delta \bar{f}}{f_0} \right) \tag{1.34}$$

where $\dfrac{dW}{dt} = \dfrac{1}{BMVA} \dfrac{d\bar{W}}{dt}$. The inertia constant $H$ of the combined generator and turbine is defined as

$$H = \frac{W_0}{BMVA} \text{s}$$

In per unit, (1.34) is rewritten in the form

$$\frac{dW}{dt} = 2H \frac{d}{dt} \Delta f \tag{1.35}$$

where $\Delta f$ is the frequency change in pu. Dividing both sides of (1.30) by the BMVA and using (1.35), we get the pu power balance equation

$$\Delta P_m - \Delta P_L = 2H \frac{d}{dt} \Delta f + D\Delta f \tag{1.36}$$

where $D = \bar{D} f_0 / BMVA$ is the pu load damping constant. The transfer function representation of (1.36) is shown in Figure 1.12.

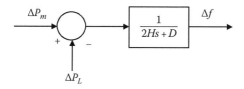

**FIGURE 1.12**
Block diagram of a generator–load model.

The following example examines the effect of the primary control (governor) on the transient and steady-state frequency when an isolated area is subject to sudden load variation.

### Example 1.4

An isolated power system consists of five identical generating units. The rating of each generator is 600 MVA and the inertia constant $H$ is 6.0 s on 600 MVA base. The total load is 1500 MW and the droop regulation constant $R$ is 2.5 Hz/pu. The operating frequency is 50 Hz. On a 3000 MVA base,

1. Find the equivalent inertia constant
2. Find the damping factor in pu
3. Draw the block diagram of the system with governor and non-reheat turbine

When a sudden drop in the load of 75 MW occurs in the system,

4. Find the transient and steady-state frequency response without the governor action
5. Find the transient and steady-state response in the presence of the governor and turbine actions. Assume the time constant of the governor is 0.2 s and the turbine is of non-reheat type with a time constant $T_t$ of 0.3 s.

**Solution:**

1. The equivalent generator has an inertia constant $H_{eq}$ equal to the sum of the individual inertia constants. Therefore, $H_{eq} = 5 \times 6 \times 600/3000 = 6$ s.
2. Assuming that the load frequency characteristic is linear, $\bar{D} = \dfrac{\bar{P}_L}{\bar{f}} = \dfrac{1500}{50} = 30$ MW/Hz. Then, the damping constant in pu

   is calculated as $D = \dfrac{\bar{D}f_0}{BMVA} = \dfrac{30*50}{3000} = 0.5$ pu.
3. The block diagram is shown in Figure 1.13.
4. Without the governor action, that is, $\Delta P_m = 0.0$, the transfer function between the frequency change as an output and the load change $\Delta P_L$ as an input is given as

$$\frac{\Delta f}{\Delta P_L} = -\frac{1}{2H_{eq}s + D}$$

For $\Delta P_L = -\dfrac{75}{3000} = -0.025$ pu, the frequency deviation $\Delta f$ in the s-domain takes the form

$$\Delta f(s) = -\frac{1}{2H_{eq}s + D}\Delta P_L(s)$$

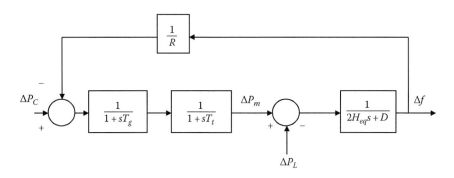

**FIGURE 1.13**
Block diagram of an isolated control area.

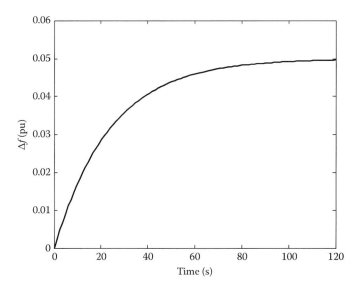

**FIGURE 1.14**
Step response of an isolated area without primary control.

Therefore,

$$\Delta f(s) = \frac{0.025}{s(12s + 0.5)}$$

The steady-state frequency deviation can be found using the final value theorem as follows: $\Delta f_{st} = \lim_{t \to \infty} \Delta f = \lim_{s \to 0} s\Delta f(s) = \frac{0.025}{0.5} = 0.05$ pu. The transient response of $\Delta f(t)$ is shown in Figure 1.14.

5. In the presence of governor-turbine actions, and assuming zero change in the speed changer ($\Delta P_c = 0$), the transfer function can be written as

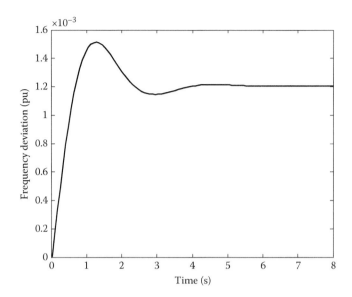

**FIGURE 1.15**
Step response of an isolated area with the primary control action.

$$\frac{\Delta f}{\Delta P_L} = -\frac{\left(1+sT_g\right)\left(1+sT_t\right)}{\left(2H_{eq}s+D\right)\left(1+sT_g\right)\left(1+sT_t\right)+\dfrac{1}{R}} \qquad (1.37)$$

From which the steady-state frequency deviation $\Delta f_{st}$ can be found as

$$\Delta f_{st} = -\frac{\Delta P_L}{\beta} \qquad (1.38)$$

where $\beta = D + \dfrac{1}{R}$ and is termed as the area frequency response characteristic. In this case, the transient response of the frequency deviation is
shown in Figure 1.15 and $\Delta f_{st} = 1.2 \times 10^{-3}$ pu.

### 1.3.3.2 Dynamic Model of Interconnected Control Areas

When two control areas are connected through a tie-line of reactance $X_{12}$, the
tie-line power flow between them, $P_{tie12}$, can be written in the form [6]

$$P_{tie12} = P_{max} \sin\left(\delta_1^0 - \delta_2^0\right) \qquad (1.39)$$

where
  $P_{max}$ is the static transmission capacity of the tie-line
  $\delta_1^0$ and $\delta_2^0$ are the nominal phase angles of the voltages at both ends

The incremental change in tie-line power $\Delta P_{tie12}$ can be obtained by linearizing (1.39) about the nominal phase angles. Then, $\Delta P_{tie12}$ takes the form

$$\Delta P_{tie12} = T_{12}\left(\Delta\delta_1 - \Delta\delta_2\right) \tag{1.40}$$

where $T_{12} = P_{\max}\cos\left(\delta_1^0 - \delta_2^0\right)$ is known as the synchronizing power coefficient. Equation 1.40 can be written in terms of the frequency deviations in both areas by using the relation $\Delta f = \dfrac{1}{2\pi}\dfrac{d\Delta\delta}{dt}$. As a result, (1.40) becomes

$$\Delta P_{tie12} = 2\pi T_{12}\left(\int \Delta f_1 dt - \int \Delta f_2\, dt\right) \tag{1.41}$$

In this case, the power balance equation of each area takes the following form:

*Area 1:*

$$\Delta P_{m1} - \Delta P_{L1} = 2H_1\frac{d}{dt}\Delta f_1 + D_1\Delta f_1 + \Delta P_{tie12} \tag{1.42}$$

*Area 2:*

$$\Delta P_{m2} - \Delta P_{L2} = 2H_2\frac{d}{dt}\Delta f_2 + D_2\Delta f_2 + \Delta P_{tie21} \tag{1.43}$$

where $\Delta P_{tie21} = -\Delta P_{tie12}$.

Note that, the Laplace transformed forms of (1.41) through (1.43) become

$$\Delta P_{tie12} = \frac{2\pi T_{12}}{s}\left(\Delta f_1 - \Delta f_2\right) \tag{1.44}$$

$$\Delta P_{m1} - \Delta P_{L1} = \left(2H_1 s + D_1\right)\Delta f_1 + \Delta P_{tie12} \tag{1.45}$$

$$\Delta P_{m2} - \Delta P_{L2} = \left(2H_2 s + D_2\right)\Delta f_2 - \Delta P_{tie12} \tag{1.46}$$

These three equations are used to construct the block diagram of a two-area power system under the primary control actions, as shown in Figure 1.16.

The steady-state frequency deviation for the system shown in Figure 1.16 can be determined as follows. Consider load changes (step changes) $\Delta P_{L1}$ and $\Delta P_{L2}$ taking place in area 1 and area 2, respectively. At steady state, the frequency deviations in both areas will be the same. Therefore,

$$\Delta f_1 = \Delta f_2 = \Delta f \tag{1.47}$$

The incremental mechanical power changes $\Delta P_{m1}$ and $\Delta P_{m2}$ at steady state is determined through the speed regulation constant as

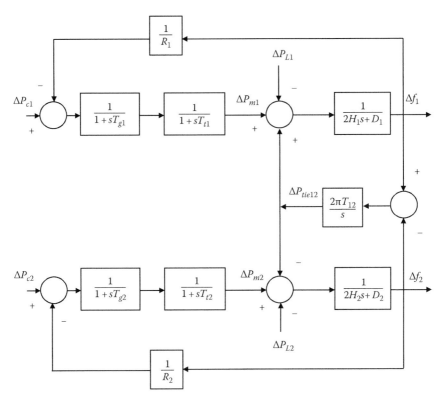

**FIGURE 1.16**
Block diagram of a two-area power system under the primary control actions.

$$\Delta P_{m1} = -\frac{\Delta f}{R_1} \tag{1.48}$$

$$\Delta P_{m2} = -\frac{\Delta f}{R_2} \tag{1.49}$$

Using (1.48) and (1.49), we find the steady-state forms of (1.45) and (1.46) as

$$-\frac{\Delta f}{R_1} - \Delta P_{L1} - \Delta P_{tie12} = D_1 \Delta f \tag{1.50}$$

$$-\frac{\Delta f}{R_2} - \Delta P_{L2} + \Delta P_{tie12} = D_2 \Delta f \tag{1.51}$$

By adding (1.50) and (1.51), the steady-state frequency deviation takes the form

$$\Delta f = -\frac{\left(\Delta P_{L1} + \Delta P_{L2}\right)}{D_1 + D_2 + \dfrac{1}{R_1} + \dfrac{1}{R_2}} \tag{1.52}$$

The steady-state tie-line power deviation can be determined from (1.51) as

$$\Delta P_{tie12} = \left(D_2 + \frac{1}{R_2}\right)\Delta f + \Delta P_{L2} \tag{1.53}$$

Substituting (1.52) into (1.53), we get

$$\Delta P_{tie12} = \frac{-\left(D_2 + \dfrac{1}{R_2}\right)\Delta P_{L1} + \left(D_1 + \dfrac{1}{R_1}\right)\Delta P_{L2}}{D_1 + D_2 + \dfrac{1}{R_1} + \dfrac{1}{R_2}} \tag{1.54}$$

In terms of the frequency response characteristics of each area $\beta_1 = D_1 + \dfrac{1}{R_1}$ and $\beta_2 = D_2 + \dfrac{1}{R_2}$, Equations 1.52 and 1.54 can be put in the forms

$$\Delta f = -\frac{\left(\Delta P_{L1} + \Delta P_{L2}\right)}{\beta_1 + \beta_2} \tag{1.55}$$

$$\Delta P_{tie12} = \frac{-\beta_2 \Delta P_{L1} + \beta_1 \Delta P_{L2}}{\beta_1 + \beta_2} \tag{1.56}$$

If the two areas are identical ($\beta_1 = \beta_2 = \beta$) and there is no load change in area 2 ($\Delta P_{L2} = 0$), then

$$\Delta f = -\frac{\Delta P_{L1}}{2\beta} \tag{1.57}$$

Comparing (1.38) and (1.57), one can conclude that the steady-state frequency deviation for the isolated area is double the value for the two-area system. Moreover, the tie-line power change will be

$$\Delta P_{tie12} = \frac{-\beta \Delta P_{L1}}{2\beta} = \frac{-\Delta P_{L1}}{2} \tag{1.58}$$

This means that the load change occurring in area 1 will be shared equally between area 1 and area 2. Therefore, in terms of LFC, it is advantageous to operate the power system as an interconnected multi-area rather than as isolated areas.

## 1.4 Automatic Generation Control of Interconnected Power Areas

As shown in Example 1.4, the mere presence of the primary control via governor action is not enough to maintain zero frequency deviation when the system has experienced a load change. In order to reduce the frequency change to zero with acceptable transient performance, another control action should be provided. Hence, the restoration of the system frequency to nominal value with desired transient specifications requires a secondary control loop that adjusts the load reference set point $\Delta P_c$ of selected generating units. Since the system load is changing continually, it is required that the generators' outputs change automatically. The main tasks of the automatic generation control (AGC) are to regulate the system frequency to the nominal value and to keep the tie-line power interchange between different areas at the scheduled levels by adjusting the output of the selected generators.

### 1.4.1 AGC in a Single-Area Power System

A secondary control loop that manipulates the output frequency deviation to produce a load reference set point can be a simple reset action (integral) or a PID (proportional integral derivative) controller. The integral secondary

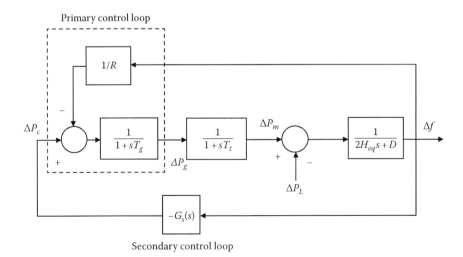

**FIGURE 1.17**
Block diagram of an isolated area equipped with primary and secondary controls.

control loop forces the steady-state frequency deviation to zero while a PID controller affects both the transient behavior and the steady-state value of the frequency deviation. Figure 1.17 shows a block diagram of an isolated area equipped with a secondary control loop, which has a transfer function $G_s(s)$. It is worth mentioning that the AGC can be installed on selected generating units to adjust their load reference settings to overcome the effects of composite frequency regulation characteristics. As a result of the presence of AGC on selected generating units, the frequency of all other units with an AGC will be restored. The following example shows the effects of different types of secondary control.

### Example 1.5

Consider the isolated system given in Example 1.4. Assume that a secondary control is installed on one of the five generators. Investigate the frequency deviation response for a load change of 20% when the secondary control is

1. A pure integral control in the form $G_s(s) = \dfrac{K_I}{s}$ with $K_I = 10$

2. A proportional integral (PI) control in the form $G_s(s) = \left( K_P + \dfrac{K_I}{s} \right)$

   where $K_P = K_I = 10$

3. A PID control of the form $G_s(s) = \left( K_P + \dfrac{K_I}{s} + K_D s \right)$ with

   $K_P = K_I = 10$ and $K_D = 5$

### Solution:

The transfer function of the system shown in Figure 1.17 with secondary controller $G_s(s)$ can be written as

$$\frac{\Delta f}{\Delta P_L} = -\frac{\left(1 + sT_g\right)\left(1 + sT_t\right)}{\left(2H_{eq}s + D\right)\left(1 + sT_g\right)\left(1 + sT_t\right) + \left(\dfrac{1}{R} + G_s(s)\right)}$$

Step response for the three different types of $G_s(s)$ is shown in Figure 1.18. It is clear that

1. The integral control action of the three controllers is capable to restore the frequency to the nominal value
2. The effect of the derivative term of the PID is to damp out the frequency deviations to a desired level

A systematic design procedure for different types of secondary loop controllers will be covered in subsequent chapters.

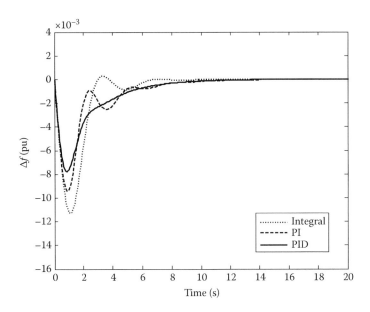

**FIGURE 1.18**
Time response of frequency deviation for different types of secondary loop controllers.

### 1.4.2 AGC in Interconnected Power Systems

The fundamental objective of secondary control is to reestablish the balance between the area load and its generation. This can be achieved when the control action keeps the frequency deviation close to zero and the net interchange power with neighboring areas at scheduled levels. A control signal of the secondary control loop known as the area control error (ACE) is used to realize the desired objectives. In the literature, there are many forms of ACE. The most commonly used form of ACE consists of two terms, one is the tie-line power flow deviation $\Delta P_{tie12}$ added to a second term, which is proportional to the frequency deviation $\Delta f$. Thus, the ACE signals for a two-area system are given by

$$ACE_1 = \Delta P_{tie12} + B_1\Delta f_1 \tag{1.59}$$

$$ACE_2 = \Delta P_{tie21} + B_2\Delta f_2 \tag{1.60}$$

where $B_1$ and $B_2$ are termed as the bias factors. For two identical areas with $\Delta P_{L2}=0$, if we substitute (1.57) and (1.58) in (1.59) and (1.60), we get

$$ACE_1 = \Delta P_{tie12} + B_1\Delta f_1 = \frac{-\Delta P_{L1}}{2} - B_1\frac{\Delta P_{L1}}{2\beta} \tag{1.61}$$

$$ACE_2 = \Delta P_{tie21} + B_2 \Delta f_2 = \frac{\Delta P_{L1}}{2} - B_2 \frac{\Delta P_{L1}}{2\beta} \tag{1.62}$$

Ideally, the task of secondary control of a given area is to correct the changes occurring in that area. In order to achieve that, the bias factor $B_1 = B_2 = B$ (identical areas) is selected equal to the frequency response characteristic $\beta$, that is, $B = \beta$. In this case,

$$ACE_1 = -\Delta P_{L1} \tag{1.63}$$

$$ACE_2 = 0.0 \tag{1.64}$$

Equations 1.63 and 1.64 indicate that the secondary control of area 1 will respond to its load change $\Delta P_{L1}$ and change its generation in order to bring the $ACE_1$ to zero, and the secondary control of area 2 has no effect on area 1. Therefore, the logic choice for bias factors are $B_1 = \beta_1 = D_1 + \frac{1}{R_1}$ and $B_2 = \beta_2 = D_2 + \frac{1}{R_2}$. A block diagram of a two-area system with secondary control is depicted in Figure 1.19.

### Example 1.6

Consider a two-area power system with the following data: Speed regulation $R_1 = 0.05, R_2 = 0.07$, damping factor $D_1 = 1, D_2 = 0.8$, inertia constant $H_1 = 6, H_2 = 4$, governor time constant $T_{g1} = 0.25, T_{g2} = 0.35$, non-reheat turbine time constant $T_{t1} = 0.5, T_{t2} = 0.6, BMVA_1 = BMVA_2 = 1000$, and the nominal frequency is 60 Hz. The synchronizing power coefficient $2\pi T_{12} = 2$. Area 1 and area 2 are subject to load disturbances of 250 and 100 MW, respectively:

a. Find the steady-state frequency and tie-line power deviations if no AGC is installed. Find the mechanical power change of each area. Explain the obtained results.
b. If an AGC is installed in each area, as shown in Figure 1.19, find the transfer function matrix of the complete system.
c. Simulate the complete system with $G_{s1}(s) = \frac{K_{I1}}{s}$ and $G_{s2}(s) = \frac{K_{I2}}{s}$, where $K_{I1} = K_{I2} = 0.2$.

### Solution:

a. The bias factors are calculated as $B_1 = \beta_1 = D_1 + \frac{1}{R_1} = 21$ and $B_2 = \beta_2 = D_2 + \frac{1}{R_2} = 15.086$. With no AGC in both areas, use (1.52) and (1.54) to get the steady-state deviations in frequency and tie-line power as $\Delta f = -\frac{0.25 + 0.1}{21 + 15.086} = -0.0097\,\text{pu} = 0.582\,\text{Hz}$ and

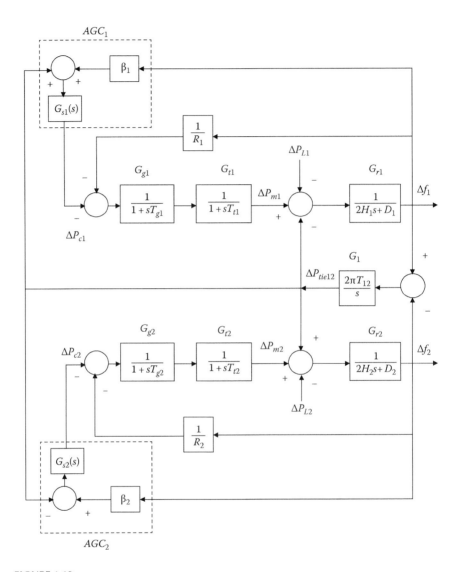

**FIGURE 1.19**
Block diagram of a two-area power system with AGC control.

$$\Delta P_{tie12} = \frac{-0.25 \times 15.086 + 0.1 \times 21}{21 + 15.086} = -0.0463 \text{ pu} = 46.3 \text{ MW}. \quad \text{From}$$

(1.48) and (1.49), the mechanical power deviations will be

$$\Delta P_{m1} = -\frac{(-0.0097)}{0.05} = 0.194 \text{ pu} = 194 \text{ MW and } \Delta P_{m2} = -\frac{(-0.0097)}{0.07} =$$

$0.13857 \text{ pu} = 138.5 \text{ MW}$. This means that the frequency and tie-line power deviations are not vanishing due to the absence of AGC.

b. The system shown in Figure 1.19 is a two-input ($\Delta P_{L1}$ and $\Delta P_{L2}$) two-output ($\Delta f_1$ and $\Delta f_2$) system. From the block diagram in Figure 1.19, one can write the following equations:

$$\Delta P_{tie12} = G_1\left(\Delta f_1 - \Delta f_2\right) \tag{1.65}$$

$$\left(\Delta P_{m1} - \Delta P_{L1} - \Delta P_{tie12}\right)G_{r1} = \Delta f_1 \tag{1.66}$$

$$\Delta P_{m1} = \left(-\frac{\Delta f_1}{R_1} - G_{s1}\left(\beta_1\Delta f_1 + \Delta P_{tie12}\right)\right)G_{g1}G_{t1} \tag{1.67}$$

$$\left(\Delta P_{m2} - \Delta P_{L2} + \Delta P_{tie12}\right)G_{r2} = \Delta f_2 \tag{1.68}$$

$$\Delta P_{m2} = \left(-\frac{\Delta f_2}{R_2} - G_{s2}\left(\beta_2\Delta f_2 - \Delta P_{tie12}\right)\right)G_{g2}G_{t2} \tag{1.69}$$

where

$$G_1 = \frac{2\pi T_{12}}{s}$$

$$G_{r1} = \frac{1}{\left(2H_1 s + D_1\right)}$$

$$G_{r2} = \frac{1}{\left(2H_2 s + D_2\right)}$$

$$G_{g1} = \frac{1}{\left(2H_2 s + D_2\right)}$$

Using (1.65) and (1.67) into (1.66), we obtain the following equation:

$$G_{11}\Delta f_1 + G_{12}\Delta f_2 = -G_{r1}\Delta P_{L1} \tag{1.70}$$

where

$$G_{11} = 1 + G_{g1}G_{t1}G_{r1}\left(\frac{1}{R_1} + \beta_1 G_{s1} + G_1 G_{s1}\right) + G_{r1}G_1 \tag{1.71}$$

$$G_{12} = -G_1 G_{r1}\left(1 + G_{g1}G_{t1}G_{s1}\right) \tag{1.72}$$

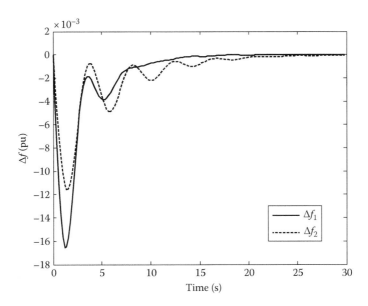

**FIGURE 1.20**
Frequency changes of Example 1.6.

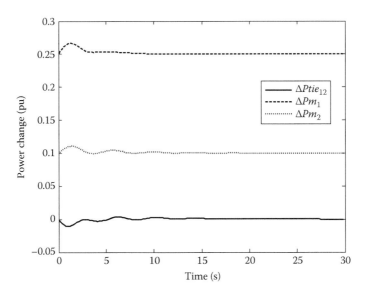

**FIGURE 1.21**
Mechanical power and tie-line power changes of Example 1.6.

Similarly, upon substitution of (1.65) and (1.69) in (1.68), we get

$$G_{21}\Delta f_1 + G_{22}\Delta f_2 = -G_{r2}\Delta P_{L2} \tag{1.73}$$

where

$$G_{21} = -G_1 G_{r2} \left(1 + G_{g2} G_{t2} G_{s2}\right) \tag{1.74}$$

$$G_{22} = 1 + G_{g2} G_{t2} G_{r2} \left(\frac{1}{R_2} + \beta_2 G_{s2} + G_1 G_{s1}\right) + G_{r2} G_1 \tag{1.75}$$

Equations 1.70 and 1.73 can be put in the following matrix form:

$$\begin{bmatrix} \Delta f_1 \\ \Delta f_2 \end{bmatrix} = \begin{bmatrix} G_{11} & G_{12} \\ G_{21} & G_{22} \end{bmatrix}^{-1} \begin{bmatrix} -G_{r1} & 0 \\ 0 & -G_{r2} \end{bmatrix} \begin{bmatrix} \Delta P_{L1} \\ \Delta P_{L2} \end{bmatrix} = M \begin{bmatrix} \Delta P_{L1} \\ \Delta P_{L2} \end{bmatrix} \tag{1.76}$$

where $M$ is the transfer function matrix
c. The MATLAB is used to find the transient frequency deviations of each area using (1.76). The results are shown in Figure 1.20. The mechanical power changes as well as the tie-line power are given in Figure 1.21.

It is evident that the AGC is capable of restoring the frequency of both areas to the nominal value while maintaining the tie-line power at the scheduled value.

# References

1. O. Elgerd, *Electric Energy Systems Theory: An Introduction*, McGraw-Hill, New York, 1982.
2. P. Kundur, *Power System Stability and Control*, McGraw-Hill, New York, 1994.
3. IEEE Power and Energy Society, Dynamic models for turbine-governors in power system studies, Technical report, PES-TR1, The Institute of Electrical and Electronic Engineers, Inc., Piscataway, NJ. January 2013.
4. CIGRE, Modeling of gas turbines and steam turbines in combined-cycle power plants, Technical Brochure 238, www.e-cigre.org, December 2003.
5. A. J. Wood and B. F. Wollenberg, *Power Generation, Operation and Control*, 2nd edn., John Wiley & Sons, New York, 1996.
6. H. Saadat, *Power System Analysis*, 2nd edn., McGraw Hill, New York, 2002.

# 2

## Modeling of Multi-Area Power Systems

## 2.1 Introduction

In this chapter, we present a mathematical model for multi-area power systems using the state variable approach. We start by writing the state variable model for a single area power system and a two-area power system, and then generalize it for a multi-area system. Pole placement controller and pole placement plus integral controller will be developed for isolated load frequency control (LFC) area. The design of an observer-based state feedback plus integral control observer is presented. Optimal control static and dynamic output feedback controllers of multi-area power system are given at the end of this chapter.

## 2.2 State Space Model of an Isolated Power Area

Consider the block diagram of an isolated area equipped with primary and secondary control loops, as shown in Figure 1.17. In order to develop the state variable model, we define the following state vector $x = \begin{bmatrix} x_1 & x_2 & x_3 \end{bmatrix}^T$ where $x_1 = \Delta f$, $x_2 = \Delta P_m$ and $x_3 = \Delta P_g$. The transfer function

$$G_r = \frac{\Delta f}{\Delta P_m - \Delta P_L} = \frac{1}{2Hs + D}$$

can be written as the following differential equation

$$\dot{x}_1 = -\frac{D}{2H} x_1 + \frac{1}{2H} x_2 - \frac{1}{2H} \Delta P_L \qquad (2.1)$$

Similarly, the differential equations of the transfer functions

$$G_t = \frac{\Delta P_m}{\Delta P_g} = \frac{1}{T_t s + 1}$$

and

$$G_g = \frac{\Delta P_g}{\Delta P_c - \frac{\Delta f}{R}} = \frac{1}{T_g s + 1}$$

take the following state forms

$$\dot{x}_2 = -\frac{1}{T_t} x_2 + \frac{1}{T_t} x_3 \qquad (2.2)$$

$$\dot{x}_3 = -\frac{1}{RT_g} x_1 - \frac{1}{T_g} x_3 + \frac{1}{T_g} \Delta P_c \qquad (2.3)$$

In the standard compact form, (2.1) and (2.3) can be rewritten as

$$\dot{x} = Ax + Bu + Fd \qquad (2.4)$$

$$y = Cx \qquad (2.5)$$

where
$$A = \begin{bmatrix} -\dfrac{D}{2H} & \dfrac{1}{2H} & 0 \\ 0 & -\dfrac{1}{T_t} & \dfrac{1}{T_t} \\ -\dfrac{1}{RT_g} & 0 & -\dfrac{1}{T_g} \end{bmatrix}, \quad B^T = \begin{bmatrix} 0 & 0 & \dfrac{1}{T_g} \end{bmatrix}, \quad F^T = \begin{bmatrix} -\dfrac{1}{2H} & 0 & 0 \end{bmatrix},$$

$C = \begin{bmatrix} 1 & 0 & 0 \end{bmatrix}$, $u = \Delta P_c$ is the control signal and $d = \Delta P_L$ is the load disturbance.

A state feedback pole placement controller is commonly used to place the eigenvalues of the systems (2.4) and (2.5) in desired locations [1]. In what follows, we present a brief introduction to pole placement controller design for isolated LFC areas.

## 2.3 Pole Placement Controller

The pole placement controller is given as a state feedback

$$u = -Kx \tag{2.6}$$

Based on (2.6), the closed-loop system takes the form

$$\dot{x} = (A - BK)x + Fd \tag{2.7}$$

If (2.7) is controllable (if the controllability matrix $C = \begin{bmatrix} B & AB & A^2B \end{bmatrix}$ has full rank, then the system is controllable), then by appropriately designing matrix $K$, the poles of the closed-loop matrix $(A - BK)$ can be placed at any desired locations in the left-hand side of s-plane. The pole placement controller can be easily determined by first choosing the desired locations of the closed-loop poles and then solving for the feedback matrix $K$. The MATLAB® command *place* is used in the next example to design such a controller.

### Example 2.1

Consider the isolated area given in Example 1.4. Design a state feedback controller such that the dominant closed-loop poles have a damping ratio of 0.35 and a settling time of 4.0 s. Simulate the controlled system when a step load change of 0.2 pu occurs.

### Solution:

Using the numerical values given in Example 1.4, the system equation (2.4) can be written as

$$\dot{x} = \begin{bmatrix} -0.0625 & 0.083 & 0 \\ 0 & -3.33 & 3.33 \\ -100 & 0 & -5 \end{bmatrix} + \begin{bmatrix} 0 \\ 0 \\ 5 \end{bmatrix} u + \begin{bmatrix} -0.083 \\ 0 \\ 0 \end{bmatrix} d$$

The dominant closed-loop poles in relation with transient specifications are given as $P = \left( -\dfrac{4}{T_s} \pm j\omega_n\sqrt{1 - \zeta^2} \right)$, where $\xi$ is the damping ratio, $\omega_n$ is the natural frequency, and $T_s = \dfrac{4}{\zeta\omega_n}$ is the settling time. The third closed-loop pole is chosen arbitrarily 5–10 times farther to the left of the real part of the dominant poles. Therefore, the desired poles are given by $P_c = \left( \left( -\dfrac{4}{T_s} \pm j\omega_n\sqrt{1 - \zeta^2} \right), -5\left( \dfrac{4}{T_s} \right) \right) = (-1 \pm j2.67, -5)$. A MATLAB file is

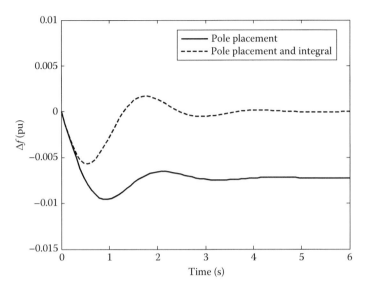

**FIGURE 2.1**
Comparison between pole placement and pole placement plus integral controllers.

prepared to design the pole placement controller. The script of this file is shown in Appendix A. The state feedback gain matrix is found to be $K = \begin{bmatrix} 8.6 & 0.343 & -0.28 \end{bmatrix}$. The closed-loop system response due to step load change of 0.2 pu is depicted in Figure 2.1. It is clear that this controller is capable of achieving the desired transient response but with a nonzero steady-state frequency deviation. In order to make this deviation zero, an integral control part is augmented to the state feedback controller, as will be shown in the next section.

## 2.4  State Feedback Control with Integral Action

An integral control action is added to the pole placement controller in order to achieve both the desired transient specifications and steady-state error. To show this, consider an additional path from the integration of the system output to the input, as shown in Figure 2.2.

The control signal of this system takes the form

$$u = -Kx - K_I v = -\begin{bmatrix} K & K_I \end{bmatrix} \begin{bmatrix} x \\ v \end{bmatrix} \qquad (2.8)$$

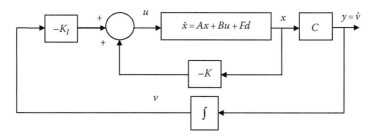

**FIGURE 2.2**
State feedback and integral control.

where

$$v = \int y\,dt \qquad (2.9)$$

We write the augmented system from (2.4) and (2.9) as

$$\begin{bmatrix} \dot{x} \\ \dot{v} \end{bmatrix} = \begin{bmatrix} A & 0 \\ C & 0 \end{bmatrix}\begin{bmatrix} x \\ v \end{bmatrix} + \begin{bmatrix} B \\ 0 \end{bmatrix}u + \begin{bmatrix} F \\ 0 \end{bmatrix}d \qquad (2.10)$$

$$y = \begin{bmatrix} C & 0 \end{bmatrix}\begin{bmatrix} x \\ v \end{bmatrix} \qquad (2.11)$$

The closed-loop augmented system is determined by substituting (2.8) in (2.10). This results in

$$\begin{bmatrix} \dot{x} \\ \dot{v} \end{bmatrix} = \begin{bmatrix} A - BK & -BK_I \\ C & 0 \end{bmatrix}\begin{bmatrix} x \\ v \end{bmatrix} + \begin{bmatrix} F \\ 0 \end{bmatrix}d \qquad (2.12)$$

Equations 2.11 and 2.12 can be put in the following compact form:

$$\dot{\bar{X}} = \left(\bar{A} - \bar{B}\bar{K}\right)\bar{X} + \bar{F}d \qquad (2.13)$$

$$y = \bar{C}\bar{X} \qquad (2.14)$$

where $\bar{X} = \begin{bmatrix} x \\ v \end{bmatrix}$, $\bar{A} = \begin{bmatrix} A & 0 \\ C & 0 \end{bmatrix}$, $\bar{B} = \begin{bmatrix} B \\ 0 \end{bmatrix}$, $\bar{F} = \begin{bmatrix} F \\ 0 \end{bmatrix}$, $\bar{C} = \begin{bmatrix} C & 0 \end{bmatrix}$, and $\bar{K} = \begin{bmatrix} K & K_I \end{bmatrix}$.

The state feedback control with integral action can be designed for the augmented system (2.13) by selecting desired poles of the closed-loop matrix $\left(\bar{A} - \bar{B}\bar{K}\right)$ and then using the MATLAB function *place* to determine the feedback gain matrix $\bar{K}$ as explained in the next example.

**Example 2.2**

For Example 2.1, design a state feedback control with an integral action to achieve the same transient response specifications ($T_s = 4$ and $\zeta = 0.35$) and to reduce the steady state error to zero.

**Solution:**

The augmented system (2.12) is of fourth order and the closed-loop poles are selected as $(-1 \pm j2.67, -5, -10)$. Running the MATLAB script given in Appendix A gives the augmented control gain matrix $\bar{K} = \begin{bmatrix} 136.24 & 2.5 & 1.72 & -293.87 \end{bmatrix}$. The step response for a load disturbance of 0.2 pu is shown in Figure 2.1. The effect of the integral action in eliminating the steady-state error and achieving the given transient specifications is evident.

## 2.5 State Estimation

Implementation of state feedback controllers requires measuring the state variables, which may not be accessible for measurement. In this case, state variables are estimated. A state estimator (observer) can be designed if a given system is observable (if the observability matrix $O^T = \begin{bmatrix} C & CA & CA^2 \end{bmatrix}$ has full rank, then the system is observable). The structure of a full-order state estimator is shown in Figure 2.3. The observer equation is given by

$$\dot{\hat{x}} = A\hat{x} + Bu + L\left(y - C\hat{x}\right) \tag{2.15}$$

or equivalently

$$\dot{\hat{x}} = \left(A - LC\right)\hat{x} + Bu + Ly \tag{2.16}$$

where $\hat{x}$ is the estimated state vector. The estimation error is defined as $e = x - \hat{x}$ and the error dynamic equation is given by

$$\dot{e} = \dot{x} - \dot{\hat{x}} \tag{2.17}$$

Substituting (2.4) and (2.15) in (2.17), we get the following estimation error dynamic (assuming $d = 0$):

$$\dot{e} = \left(A - LC\right)e \tag{2.18}$$

If the system is observable, one can choose the gain matrix $L$ to assign arbitrarily the eigenvalues of $\left(A - LC\right)$. The matrix transpose operation does

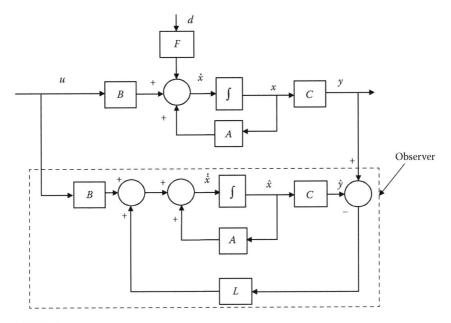

**FIGURE 2.3**
Full-order state estimator.

not affect eigenvalues, that is, the eigenvalues of $(A - LC)$ are the same as those of $(A - LC)^T = A^T - C^T L^T$. Therefore, the problem of choosing the state estimator gain vector $L$ to place the eigenvalues of $(A - LC)$ at the desired locations is equivalent to the problem of state feedback pole placement by making the following associations: $A \leftrightarrow A^T$, $B \leftrightarrow C^T$, and $K \leftrightarrow L^T$. Hence, the MATLAB function *place* can be used to find the gain vector $L$.

### Example 2.3

In this example, we design an observer for the system given in Example 2.1 in order to place the poles of the estimated states at $P_o = (-6 \pm j2, -50)$.

### Solution:

Using the associations give here along with the MATLAB function *place* as shown in Appendix A, the observer gain matrix $L$ is found to be $L = \begin{bmatrix} 53.6 & 2113.33 & 710 \end{bmatrix}^T$.

## 2.5.1 Observer-Based State Feedback with Integral Control

In Section 2.3, we presented a state feedback controller plus integral control action using pole placement. In this section, we present a design methodology that incorporates an observer to provide an estimate of the state vector for the state feedback.

The observer-based controller is of the form

$$u = -K\hat{x} - K_I v = -\begin{bmatrix} K & K_I \end{bmatrix} \begin{bmatrix} \hat{x} \\ v \end{bmatrix} \tag{2.19}$$

$$\dot{v} = y = Cx \tag{2.20}$$

$$\dot{\hat{x}} = (A - LC)\hat{x} + Bu + Ly \tag{2.21}$$

Substituting (2.19) in (2.21), we obtain

$$\dot{\hat{x}} = (A - LC - BK)\hat{x} - BK_I v + Ly \tag{2.22}$$

Equations 2.20 and 2.22 are combined together to get

$$\begin{bmatrix} \dot{\hat{x}} \\ \dot{v} \end{bmatrix} = \begin{bmatrix} A - LC - BK & -BK_I \\ 0 & 0.0 \end{bmatrix} \begin{bmatrix} \hat{x} \\ v \end{bmatrix} + \begin{bmatrix} L \\ 1 \end{bmatrix} y \tag{2.23}$$

Now substituting the control law (2.19) into (2.4), we get

$$\dot{x} = Ax - BK\hat{x} - BK_I v + Fd \tag{2.24}$$

Equations 2.23 and 2.24 form a seventh-order closed-loop state equation:

$$\begin{bmatrix} \dot{x} \\ \dot{\hat{x}} \\ \dot{v} \end{bmatrix} = \begin{bmatrix} A & -BK & -BK_I \\ LC & A - LC - BK & -BK_I \\ -C & 0 & 0 \end{bmatrix} \begin{bmatrix} x \\ \hat{x} \\ v \end{bmatrix} + \begin{bmatrix} F \\ 0 \\ 0 \end{bmatrix} d \tag{2.25}$$

with

$$y = \begin{bmatrix} C & 0 & 0 \end{bmatrix} \begin{bmatrix} x \\ \hat{x} \\ v \end{bmatrix} \tag{2.26}$$

**Example 2.4**

Design an observer-based state feedback with integral action for the system given in Example 2.1. The desired closed-loop dominant poles are located at $(-1 \pm j2.67)$ and the other two poles are assumed to be $(-5, -10)$. The observer poles are selected as five times of $(-1 \pm j2.67, -5)$. Find the state feedback gain matrix $\bar{K} = \begin{bmatrix} K & K_I \end{bmatrix}$ and the observer gain matrix $L$. Check the pole locations of the overall closed-loop system (2.25). Simulate the complete system for a step disturbance of 0.2 pu and the initial condition $x^T = \begin{bmatrix} 0.1 & 0 & 0 & 0.05 & 0.05 & 0 & 0 \end{bmatrix}$.

**Solution:**

The MATLAB script of this example is given in Appendix A. The state-feedback gain matrix is calculated as $\bar{K} = \begin{bmatrix} K & K_I \end{bmatrix} = \begin{bmatrix} 136.24 & 2.5 & 1.72 & -293.87 \end{bmatrix}$. In addition, the observer gain matrix that locates the eigenvalues of the observer error dynamics at 5 $(-1 \pm j2.67, -5)$ is found as $L = \begin{bmatrix} 27 & 2582 & 12794 \end{bmatrix}^T$. The resulting observer-based closed-loop system equation (2.25) becomes

$$
\begin{bmatrix} \dot{x}_1 \\ \dot{x}_2 \\ \dot{x}_3 \\ \dot{\hat{x}}_1 \\ \dot{\hat{x}}_2 \\ \dot{\hat{x}}_3 \\ \dot{v} \end{bmatrix}
=
\begin{bmatrix}
-0.06 & 0.08 & 0 & 0 & 0 & 0 & 0 \\
0 & -3.33 & 3.33 & 0 & 0 & 0 & 0 \\
-100 & 0 & -5 & -681.21 & -12.52 & -8.6 & 1469.38 \\
26.6 & 0 & 0 & -26.66 & 0.08 & 0 & 0 \\
2582.3 & 0 & 0 & -2582.3 & -3.33 & 3.33 & 0 \\
1279.38 & 0 & 0 & -1357.51 & -12.52 & -13.6 & -1469.38 \\
1 & 0 & 0 & 0 & 0 & 0 & 0
\end{bmatrix}
$$
$$
\times
\begin{bmatrix} x_1 \\ x_2 \\ x_3 \\ \hat{x}_1 \\ \hat{x}_2 \\ \hat{x}_3 \\ v \end{bmatrix}
+
\begin{bmatrix} -0.017 \\ 0 \\ 0 \\ 0 \\ 0 \\ 0 \\ 0 \end{bmatrix} d
\qquad (2.27)
$$

The closed-loop poles of (2.27) are calculated as $P = (-1 \pm j2.67, -5, -10, -5 \pm j13.82, -25)$, which are the same as the desired set of the closed-loop poles of the observer-based pole placement controller with integral feedback. The step response of the actual and estimated frequency deviation is presented in Figure 2.4 and the other actual and estimated states are shown in Figures 2.5 and 2.6.

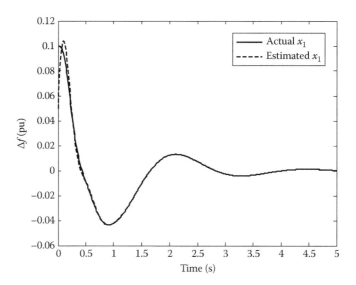

**FIGURE 2.4**
Frequency deviation $x_1$: actual and estimate.

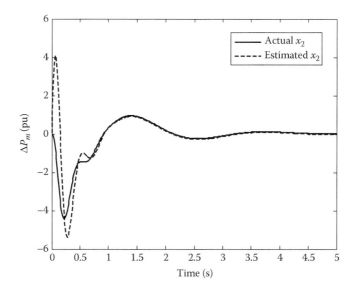

**FIGURE 2.5**
Mechanical power deviation $x_2$: actual and estimate.

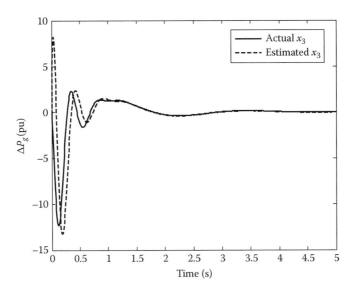

**FIGURE 2.6**
Change in governor action $x_3$: actual and estimate.

## 2.6 State Space Model of Multi-Area Power Systems

In the previous sections, we modeled a single-area load frequency scheme in state space form. In this section, a multi-area load frequency state space model will be developed. All generators belonging to the same control area will be lumped into an equivalent generating unit. To initiate this task, we start by modeling a two-area system. The model is developed by defining the state vector and the output as $x_i = \left[ \Delta f_i\ \Delta P_{mi}\ \Delta P_{gi}\ \Delta P_{tie_i} \right]^T$ and $y_i = \Delta f_i$ respectively, $i = 1,2$. Considering the change in the tie-line power defined by (1.41) as a state variable, one can write the following state equation:

$$\Delta \dot{P}_{tie1} = \Delta \dot{P}_{tie12} = 2\pi T_{12}\left( \Delta f_1 - \Delta f_2 \right) \tag{2.28}$$

$$\Delta \dot{P}_{tie2} = \Delta \dot{P}_{tie21} = 2\pi T_{21}\left( \Delta f_2 - \Delta f_1 \right) \tag{2.29}$$

where $T_{12} = T_{21}$.

The state model of the two-area system takes the following form:

$$
\begin{bmatrix} \Delta\dot{f}_1 \\ \Delta\dot{P}_{m1} \\ \Delta\dot{P}_{g1} \\ \Delta\dot{P}_{tie1} \\ \Delta\dot{f}_2 \\ \Delta\dot{P}_{m2} \\ \Delta\dot{P}_{g2} \\ \Delta\dot{P}_{tie2} \end{bmatrix}
=
\begin{bmatrix}
-D_1/2H_1 & 1/2H_1 & 0.0 & -1/2H_1 & 0.0 & 0.0 & 0.0 & 0.0 \\
0.0 & -1/T_{t_1} & 1/T_{t_1} & 0.0 & 0.0 & 0.0 & 0.0 & 0.0 \\
-1/(R_1 T_{g1}) & 0.0 & -1/T_{g1} & 0.0 & 0.0 & 0.0 & 0.0 & 0.0 \\
2\pi T_{12} & 0.0 & 0.0 & 0.0 & -2\pi T_{12} & 0.0 & 0.0 & 0.0 \\
0.0 & 0.0 & 0.0 & 0.0 & -D_2/2H_2 & 1/2H_2 & 0.0 & -1/2H_2 \\
0.0 & 0.0 & 0.0 & 0.0 & 0.0 & -1/T_{t_2} & 1/T_{t_2} & 0.0 \\
0.0 & 0.0 & 0.0 & 0.0 & -1/(R_2 T_{g2}) & 0.0 & -1/T_{g2} & 0.0 \\
-2\pi T_{12} & 0.0 & 0.0 & 0.0 & 2\pi T_{12} & 0.0 & 0.0 & 0.0
\end{bmatrix}
$$

$$
\times
\begin{bmatrix} \Delta f_1 \\ \Delta P_{m1} \\ \Delta P_{g1} \\ \Delta P_{tie1} \\ \Delta f_2 \\ \Delta P_{m2} \\ \Delta P_{g2} \\ \Delta P_{tie2} \end{bmatrix}
+
\begin{bmatrix} 0.0 & 0.0 \\ 0.0 & 0.0 \\ 1/T_{g1} & 0.0 \\ 0.0 & 0.0 \\ 0.0 & 0.0 \\ 0.0 & 0.0 \\ 0.0 & 1/T_{g2} \\ 0.0 & 0.0 \end{bmatrix}
\begin{bmatrix} u_1 \\ u_2 \end{bmatrix}
+
\begin{bmatrix} -1/2H_1 & 0.0 \\ 0.0 & 0.0 \\ 0.0 & 0.0 \\ 0.0 & 0.0 \\ 0.0 & -1/2H_2 \\ 0.0 & 0.0 \\ 0.0 & 0.0 \\ 0.0 & 0.0 \end{bmatrix}
\begin{bmatrix} d_1 \\ d_2 \end{bmatrix}
\tag{2.30}
$$

$$
\begin{bmatrix} \Delta f_1 \\ \Delta f_2 \end{bmatrix}
=
\begin{bmatrix}
1.0 & 0.0 & 0.0 & 0.0 & 0.0 & 0.0 & 0.0 & 0.0 \\
0.0 & 0.0 & 0.0 & 0.0 & 1.0 & 0.0 & 0.0 & 0.0
\end{bmatrix}
\begin{bmatrix} x_1 \\ x_2 \end{bmatrix}
\tag{2.31}
$$

where $\begin{bmatrix} x_1 \\ x_2 \end{bmatrix} = \begin{bmatrix} \Delta f_1 & \Delta P_{m1} & \Delta P_{g1} & \Delta P_{tie1} & \Delta f_2 & \Delta P_{m2} & \Delta P_{g2} & \Delta P_{tie2} \end{bmatrix}^T$.

Now, consider a multi-area power system consisting of $N$ LFC areas. The variable $\Delta P_{tie_i}$ represents the net tie-line power exchange of the $i$th area, which can be written as:

$$
\Delta\dot{P}_{tie_i} = 2\pi T_{i2}\left(\Delta f_i - \Delta f_2\right) + 2\pi T_{i3}\left(\Delta f_i - \Delta f_3\right) + \cdots 2\pi T_{iN}\left(\Delta f_i - \Delta f_N\right)
\tag{2.32}
$$

or in the equivalent form

$$
\Delta\dot{P}_{tie_i} = 2\pi\left(T_{i2} + T_{i3} + \cdots T_{iN}\right)\Delta f_i - 2\pi\left(T_{i2}\Delta f_2 + T_{i3}\Delta f_3 + \cdots T_{iN}\Delta f_N\right)
\tag{2.33}
$$

$$\Delta \dot{P}_{tie_i} = 2\pi \sum_{\substack{j=1 \\ \neq i}}^{N} T_{ij}\Delta f_i - 2\pi \sum_{\substack{j=1 \\ \neq i}}^{N} T_{ij}\Delta f_j \qquad (2.34)$$

A block diagram of the multi-area LFC system is shown in Figure 2.7.

The state space model of the $i$th area can be derived from the block diagram and it takes the following form:

$$\left.\begin{aligned} \dot{\underline{x}}_i &= \underline{A}_i\underline{x}_i + \underline{B}_i\underline{u}_i + \sum_{j=1, j\neq i}^{N} \underline{A}_{ij}\underline{x}_j + \underline{F}_i\Delta P_{d_i} \\ \underline{y}_i &= \underline{C}_i\underline{x}_i \end{aligned}\right\} \qquad (2.35)$$

where $\underline{x}_i = \left[\Delta f_i\, \Delta P_{mi}\, \Delta P_{gi}\, \Delta P_{tie_i}\right]^T$ and the model matrices are given by

$$\underline{A}_i = \begin{bmatrix} -D_i/2H_i & 1/2H_i & 0 & -1/2H_i \\ 0 & -1/T_{t_i} & 1/T_{t_i} & 0 \\ -1/(R_iT_{g_i}) & 0 & -1/T_{g_i} & 0 \\ 2\pi\sum_{j=1, j\neq i}^{N} T_{ij} & 0 & 0 & 0 \end{bmatrix}$$

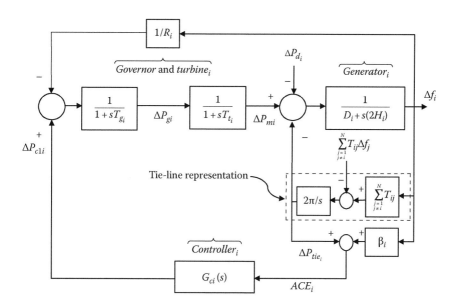

**FIGURE 2.7**

Block diagram model of the $i$th area in a multi-area power system.

$$\underline{A}_{ij} = \begin{bmatrix} 0 & 0 & 0 & 0 \\ 0 & 0 & 0 & 0 \\ 0 & 0 & 0 & 0 \\ -2\pi T_{ij} & 0 & 0 & 0 \end{bmatrix}$$

$$\underline{B}_i^T = \begin{bmatrix} 0 & 0 & \dfrac{1}{T_{gi}} & 0 \end{bmatrix}, \ \underline{F}_i^T = \begin{bmatrix} \dfrac{1}{2H_i} & 0 & 0 & 0 \end{bmatrix} \text{and } \underline{C}_i = \begin{bmatrix} 1 & 0 & 0 & 0 \end{bmatrix}$$

## 2.6.1 Including PI Controller in the Multi-Area Model [2,3]

The proportional integral (PI) controller of the form

$$G_{c_i}(s) = K_{p_i} + \frac{K_{l_i}}{s} \tag{2.36}$$

$$u_i = -K_{p_i} ACE_i - K_{l_i} \int ACE_i \, dt = -K_i y_i \tag{2.37}$$

where $K_i = \begin{bmatrix} K_{p_i} & K_{l_i} \end{bmatrix}$ can be imbedded in the multi-area model by transforming the PI control problem into an output feedback control problem. To this effect, a new state variable, $\int ACE_i \, dt$, is introduced such that the state vector of each area becomes $x_i = \begin{bmatrix} \Delta f_i & \Delta P_{mi} & \Delta P_{gi} & \Delta P_{tie_i} & \int ACE_i \, dt \end{bmatrix}^T$, where $ACE_i = \beta_i \Delta f_i + \Delta P_{tie_i}$. In this case, the output vector is redefined as $y_i = \begin{bmatrix} ACE_i & \int ACE_i \, dt \end{bmatrix}^T$ and the state model of the output feedback problem for $N$-area system can be written as

$$\left. \begin{aligned} x_i &= A_i x_i + B_i u_i + \sum_{j=1, j\neq i}^{N} A_{ij} x_j + F_i \Delta P_{d_i} \\ y_i &= C_i x_i \end{aligned} \right\} \tag{2.38}$$

with the following matrices

$$A_i = \begin{bmatrix} -D_i/2H_i & 1/2H_i & 0 & -1/2H_i & 0 \\ 0 & -1/T_{t_i} & 1/T_{t_i} & 0 & 0 \\ -1/(R_i T_{gi}) & 0 & -1/T_{gi} & 0 & 0 \\ 2\pi \sum_{j=1, j\neq i}^{N} T_{ij} & 0 & 0 & 0 & 0 \\ \beta_i & 0 & 0 & 1 & 0 \end{bmatrix}$$

$$A_{ij} = \begin{bmatrix} 0 & 0 & 0 & 0 & 0 \\ 0 & 0 & 0 & 0 & 0 \\ 0 & 0 & 0 & 0 & 0 \\ -2\pi T_{ij} & 0 & 0 & 0 & 0 \\ 0 & 0 & 0 & 0 & 0 \end{bmatrix}$$

$$B_i^T = \begin{bmatrix} 0 & 0 & 1/T_{gi} & 0 & 0 \end{bmatrix}, \quad F_i^T = \begin{bmatrix} -1/2H_i & 0 & 0 & 0 & 0 \end{bmatrix}$$

$$\text{and} \quad C_i = \begin{bmatrix} \beta_i & 0 & 0 & 1 & 0 \\ 0 & 0 & 0 & 0 & 1 \end{bmatrix}$$

Note that the output vector of (2.38) is just a virtual vector and the physically measurable output is the $ACE_i$ signal. Having introduced two different models (2.35) and (2.38) for a multi-area LFC power system, we present next two output feedback controllers using linear quadratic regulator approach, namely, the static and dynamic output feedback controllers.

## 2.7 Optimal Control-Based Output Feedback of Multi-Area Power Systems

Pole placement state feedback controllers assume that all states are available for feedback. In practice, often only the system outputs are available for feedback. Output feedback will allow us to design plant controllers of any desired structure [5]. This is another reason for preferring output feedback over state feedback. Output feedback can be classified into static and dynamic feedback [4]. To develop the output feedback controllers for a multi-area LFC system, we rewrite the models given in either (2.35) or (2.38) in the form

$$\dot{X} = AX + BU + FD \tag{2.39}$$

along with

$$Y = CX \tag{2.40}$$

where $A = \begin{bmatrix} A_1 & A_{12} \cdots A_{1N} \\ A_{21} & A_2 \cdots A_{2N} \\ \vdots & \vdots \ddots \vdots \\ A_{N1} & A_{N2} \cdots A_N \end{bmatrix}$, $B = \text{diag} \begin{bmatrix} B_1 & B_2 \cdots B_N \end{bmatrix}$, $F = \text{diag} \begin{bmatrix} F_1 & F_2 \cdots F_N \end{bmatrix}$,

$C = \text{diag} \begin{bmatrix} C_1 & C_2 \cdots C_N \end{bmatrix}$. The composite state vector $X$, the control input $U$, and the disturbance $D$ are defined by $X = \begin{bmatrix} x_1 & x_2 \cdots x_N \end{bmatrix}^T$, $U = \begin{bmatrix} u_1 & u_2 \cdots u_N \end{bmatrix}^T$, $D = \begin{bmatrix} d_1 & d_2 \cdots d_N \end{bmatrix}^T$. Note that for the model (2.35) $x_i \in \Re^4$ and $X \in \Re^{4N}$, while for the model (2.38) $x_i \in \Re^5$ and $X \in \Re^{5N}$. In what follows, we present static and output feedback control for the multi-area model given by (2.38).

### 2.7.1 Static Output Feedback

Given the linear system (2.39) and (2.40), the static output feedback takes the form

$$U = -KY \tag{2.41}$$

where $K = \text{diag} \begin{bmatrix} K_1 & K_2 \cdots K_N \end{bmatrix}$ is the feedback gain matrix with $K_i = \begin{bmatrix} K_{p_i} & K_{l_i} \end{bmatrix}$. By substituting this control law into (2.39), the following closed-loop system equation is found:

$$\dot{X} = (A - BKC) X + FD \tag{2.42}$$

$$\dot{X} = A_c X + FD \tag{2.43}$$

where $A_c = (A - BKC)$. The gain matrix can be found using a linear quadratic regulator or pole placement techniques.

### 2.7.2 Dynamic Output Feedback

A dynamic output feedback controller takes the form

$$U = WZ + VY \tag{2.44}$$

$$\dot{Z} = EZ + GY \tag{2.45}$$

where the vector $Z$ is of dimension $p$ and the matrices $W$, $V$, $E$, and $G$ are of appropriate dimensions. The matrix form of (2.44) and (2.45) is obtained as

$$\begin{bmatrix} U \\ \dot{Z} \end{bmatrix} = \begin{bmatrix} V & W \\ G & E \end{bmatrix} \begin{bmatrix} Y \\ Z \end{bmatrix} = -M \begin{bmatrix} Y \\ Z \end{bmatrix} \tag{2.46}$$

or as the following compact form

$$\hat{U} = -M\hat{Y} \tag{2.47}$$

where $\hat{U} = \begin{bmatrix} U \\ \dot{Z} \end{bmatrix}$, $\hat{Y} = \begin{bmatrix} Y \\ Z \end{bmatrix}$, and $M = -\begin{bmatrix} V & W \\ G & E \end{bmatrix}$

The idea here is to reformulate the dynamic output feedback into the form of static output feedback. To do this, the plant model (2.39) along with (2.45) can be written in the following form

$$\begin{bmatrix} \dot{X} \\ \dot{Z} \end{bmatrix} = \begin{bmatrix} A & 0 \\ 0 & 0 \end{bmatrix}\begin{bmatrix} X \\ Z \end{bmatrix} + \begin{bmatrix} B & 0 \\ 0 & I \end{bmatrix}\begin{bmatrix} U \\ \dot{Z} \end{bmatrix} + \begin{bmatrix} F & 0 \\ 0 & 0 \end{bmatrix}\begin{bmatrix} D \\ 0 \end{bmatrix} \tag{2.48}$$

Defining $W = \begin{bmatrix} X \\ Z \end{bmatrix}$, $\hat{U} = \begin{bmatrix} U \\ \dot{Z} \end{bmatrix}$, $\hat{A} = \begin{bmatrix} A & 0 \\ 0 & 0 \end{bmatrix}$, $\hat{B} = \begin{bmatrix} B & 0 \\ 0 & 1 \end{bmatrix}$, $\hat{F} = \begin{bmatrix} F & 0 \\ 0 & 0 \end{bmatrix}$, and $\hat{D} = \begin{bmatrix} D \\ 0 \end{bmatrix}$, we write (2.48) as

$$\dot{W} = \hat{A}W + \hat{B}\hat{U} + \hat{F}\hat{D} \tag{2.49}$$

The vector $\hat{Y} = \begin{bmatrix} Y \\ Z \end{bmatrix}$ can be written as

$$\hat{Y} = \begin{bmatrix} Y \\ Z \end{bmatrix} = \begin{bmatrix} C & 0 \\ 0 & I \end{bmatrix}\begin{bmatrix} X \\ Z \end{bmatrix} \tag{2.50}$$

or in terms of the vector $W$

$$\hat{Y} = \hat{C}W \tag{2.51}$$

where $\hat{C} = \begin{bmatrix} C & 0 \\ 0 & I \end{bmatrix}$.

The overall system model under dynamic feedback is now given by (2.47), (2.49) and (2.51). Substituting (2.51) in (2.47) and then substituting the result into (2.49), we get the following closed-loop system:

$$\dot{W} = \left(\hat{A} - \hat{B}M\hat{C}\right)W + \hat{F}\hat{D} \tag{2.52}$$

$$\dot{W} = \hat{A}_c W + \hat{F}\hat{D} \tag{2.53}$$

where $\hat{A}_c = \left(\hat{A} - \hat{B}M\hat{C}\right)$. A comparison of (2.52) with (2.42) shows that the dynamic output feedback has been reformulated into static output feedback. The unknown matrices $W$, $V$, $E$, and $G$ are combined into a single matrix $M$ to be designed for the $(5N + p)$-order system (2.52). The static output feedback matrix $K$ and the dynamic output feedback matrix $M$ will be determined using the optimal control technique. A brief introduction of the optimal control-based output feedback is given next.

### 2.7.3 The LQR Problem

In the regulator problem, the designer is interested in achieving good time response while guaranteeing the stability of the closed-loop system. These goals can be attained optimally by looking for the best trade-off between performance and cost of control. The objective of the state regulation of a system is to drive any initial condition to zero, thus guaranteeing stability. This may be achieved by selecting the control input $U$ to minimize a quadratic cost or performance index $J$ of the form

$$J = \frac{1}{2}\int_0^\infty \left(X^T QX + U^T RU\right) dt \tag{2.54}$$

where
  $Q$ is a positive semidefinite matrix
  $R$ is a positive definite weighting matrix

Positive definiteness (semidefiniteness) of a square matrix $S$ denoted by $S > 0\, (S \geq 0)$ is equivalent to all eigenvalues being positive (nonnegative). The terms $X^T QX$ and $U^T RU$ of the integrand are quadratic forms that measure the performance and the cost of control, respectively. The choice of the elements of $Q$ and $R$ allows the relative weighting of state variables and control inputs. Now, substituting the static output feedback control given by (2.41) into (2.54), we get

$$J = \frac{1}{2}\int_0^\infty X^T \left(Q + C^T K^T RKC\right) X\, dt \tag{2.55}$$

It can be shown that [5] if the closed-loop system is stable, that is, $X(\infty) \to 0$, then a constant, real, symmetric , and positive definite matrix $P$ exists, which satisfies

$$\frac{d}{dt}\left(X^T PX\right) = -X^T \left(Q + C^T K^T RKC\right) X \tag{2.56}$$

Substituting (2.56) into (2.55) gives

$$J = -\frac{1}{2}\int_0^\infty \frac{d}{dt}\left(X^T PX\right) dt = -\frac{1}{2}X^T PX\Big|_0^\infty \tag{2.57}$$

Since $X(\infty) \rightarrow 0$, (2.57) becomes

$$J = \frac{1}{2} X^T(0) P X(0) \tag{2.58}$$

For an undisturbed system, $D = 0$, Equation 2.46 becomes

$$\dot{X} = A_c X \tag{2.59}$$

Using (2.59), the left-hand side of (2.56) can be written as

$$\frac{d}{dt}\left(X^T P X\right) = \dot{X}^T P X + X^T P \dot{X} = X^T\left(A_c^T P + P A_c\right) X \tag{2.60}$$

Comparing the right-hand sides of (2.56) and (2.60) yields

$$A_c^T P + P A_c = -\left(Q + C^T K^T R K C\right) \tag{2.61}$$

This procedure is repeated for the dynamic output feedback system given by (2.47), (2.51), and (2.52) to obtain the following equations:

$$J_d = \frac{1}{2} \int_0^\infty W^T\left(Q_d + \hat{C}^T M^T R_d M \hat{C}\right) W \, dt \tag{2.62}$$

$$\hat{A}_c^T P_d + P_d \hat{A}_c = -\left(Q_d + \hat{C}^T M^T R_d M \hat{C}\right) \tag{2.63}$$

$$J_d = \frac{1}{2} W^T(0) P_d W(0) \tag{2.64}$$

where $Q_d$ and $R_d$ are the weighting matrices for the dynamic output feedback and $P_d$ is the solution of (2.63). Equation 2.61 (2.63) is termed as the Lyapunov matrix equation. In summary, for any fixed feedback matrix $K$ for static output feedback (or $M$ for dynamic output feedback), if a constant symmetric positive semi-definite matrix $P$ ($P_d$) exists that satisfies (2.61) (2.63) and if the closed-loop system is stable, then the cost function $J$ ($J_d$) is given by (2.58) (2.64). The optimal solution depends on the initial conditions $X(0)$ ($W(0)$).

### 2.7.4 Necessary Conditions for the Solution of the LQR Problem with Dynamic Output Feedback

The necessary conditions to solve the LQR problem with dynamic output feedback can be determined using the trace of (2.64). The trace of a matrix is defined as the sum of its diagonal elements. Let $\Phi, \Theta,$ and $\Psi$ be matrices with compatible dimensions. Some facts for the trace are

$$\text{tr}\left(\Phi\Theta\right) = \text{tr}\left(\Theta\Phi\right) \tag{2.65}$$

$$\frac{\partial}{\partial\Theta}\,\text{tr}\left(\Phi\Theta\Psi\right) = \Phi^T\Psi^T \tag{2.66}$$

$$\frac{\partial}{\partial\Theta}\,\text{tr}\left(\Phi\Theta^T\Psi\right) = \Psi\Phi \tag{2.67}$$

Taking the trace of (2.64), we get

$$J_d = \frac{1}{2}\,\text{tr}\left(W^T\left(0\right)P_d W\left(0\right)\right) = \frac{1}{2}\,\text{tr}\left(P_d W\left(0\right)W^T\left(0\right)\right) = \frac{1}{2}\,\text{tr}\left(P_d\overline{W}\right) \tag{2.68}$$

where $\overline{W} = W\left(0\right)W^T\left(0\right)$.

The problem of determining the matrix $M$ to minimize (2.62) subject to (2.53) (with $\hat{D} = 0$) is equivalent to the algebraic problem of determining $M$ to minimize (2.68) subject to (2.63). This modified problem can be easily solved using the Lagrange multiplier approach. In this case, the Lagrange function $\Lambda$ is formulated by adjoining the constraint (2.63) to the performance index (2.68) as

$$\Lambda = \text{tr}\left(P_d\overline{W}\right) + \text{tr}\left(\alpha S\right) \tag{2.69}$$

with $\alpha = \hat{A}_c^T\,P_d + P_d\,\hat{A}_c + Q_d + \hat{C}^T\,M^T R_d M\hat{C} = 0$ and $S$ is a symmetric matrix of Lagrange multipliers. The necessary conditions for minimizing (2.69) are given by

$$\frac{\partial\Lambda}{\partial S} = 0 = \alpha = \hat{A}_c^T P_d + P_d\hat{A} + Q_d + \hat{C}^T M^T R_d M\hat{C} \tag{2.70}$$

$$\frac{\partial\Lambda}{\partial P_d} = 0 = \hat{A}_c S + S\hat{A}_c^T + \overline{W} \tag{2.71}$$

$$\frac{1}{2}\frac{\partial\Lambda}{\partial M} = 0 = R_d M\hat{C}S\hat{C}_c^T - \hat{B}^T P_d S\hat{C}^T \tag{2.72}$$

If $R_d$ is positive definite and $\Gamma = \hat{C}S\hat{C}^T$ is nonsingular, then the dynamic output feedback matrix $M$ can be calculated from (2.72) as

$$M = R_d^{-1}\hat{B}^T P_d S \hat{C}^T \Gamma^{-1} \tag{2.73}$$

where $P_d$ and $S$ are the solutions of the Lyapunov equations (2.70) and (2.71), respectively. An iterative procedure is presented in [6] to find the optimal value of the performance index $J_d$ (2.68) and the feedback gain matrix $M$.

The algorithm is outlined in the following steps:

1. Initialize:
   (a) Set $k = 0$.
   (b) Select a starting gain matrix $M_0$ such that the eigenvalues of $\hat{A}_c = \hat{A} - \hat{B}M_0\hat{C}$ are in the open left-hand side of the s-plane.

2. For the $k$th iteration:
   (a) Set $\hat{A}_{c_k} = \hat{A} - \hat{B}M_k\hat{C}$.
   (b) Solve the two Lyapunov equations (2.70) and (2.71).
   (c) Set $J_{d_k} = \dfrac{1}{2}\mathrm{tr}\left(P_{d_k}\overline{W}\right)$.
   (d) $\Delta M = M_k - M_0 = R_d^{-1}\hat{B}^T P_{dk}S_k\hat{C}^T\left(\hat{C}S_k\hat{C}^T\right)^{-1} - M_0$.
   (e) Update the gain matrix: $M_{k+1} = M_k + \gamma\Delta M$, where $\gamma$ is selected such that the eigenvalues of $\hat{A}_c = \hat{A} - \hat{B}M_{k+1}\hat{C}$ are in the open left-hand side of the s-plane.

3. Stopping criteria:
   (a) $J_{d_{k+1}} = \dfrac{1}{2}\mathrm{tr}\left(P_{d_{k+1}}\overline{W}\right)$.
   (b) If $\left|J_{d_{k+1}} - J_{d_k}\right| \le \varepsilon \begin{cases} \text{True} \to \text{Terminate and set } M = M_{k+1} \text{ and } J_d = J_{dk+1}. \\ \text{False} \to \text{set } k = k+1 \text{ and go to 2.} \end{cases}$

A block diagram implementation of the dynamic output feedback controller for a two-area LFC system is shown in Figure 2.8.

The next example illustrates the design of a dynamic output feedback via LQR for a two-area power system.

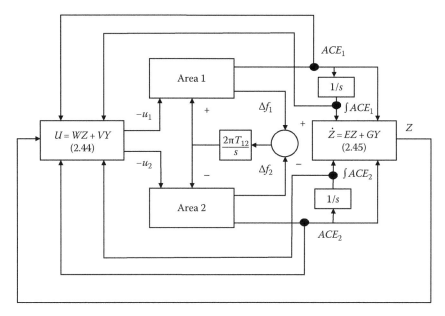

**FIGURE 2.8**
Dynamic output feedback of a two-area power system.

**TABLE 2.1**

Parameters of a two-area system

| Area # 1 | | | | | | Area # 2 | | | | | |
|---|---|---|---|---|---|---|---|---|---|---|---|
| $T_{g_1}$ | $T_{t_1}$ | $H_1$ | $D_1$ | $R_1$ | $\beta_1$ | $T_{g_2}$ | $T_{t_2}$ | $H_2$ | $D_2$ | $R_2$ | $\beta_2$ |
| 0.2 | 0.5 | 5 | 0.6 | 0.05 | 20.6 | 0.3 | 0.6 | 4 | 0.9 | 0.0625 | 16.9 |

**Example 2.5**

Consider a two-area power system with the parameters shown in Table 2.1 and the synchronizing power coefficient $T_{12} = T_{21} = \dfrac{1}{\pi}$.

1. Design an LQR with a dynamic output feedback where the system outputs are the frequency deviations.
2. Show the transient behavior of the frequency, mechanical, and tie-line power deviations when the system is subject to initial frequency deviations $\Delta f_1 = \Delta f_2 = 0.05$.

**Solution:**

In this case, the two-area model is given by (2.30) and (2.31) with the following matrices:

$$
A = \begin{bmatrix}
-0.06 & 0.1 & 0.0 & -0.1 & 0.0 & 0.0 & 0.0 & 0.0 \\
0.0 & -2.0 & 2.0 & 0.0 & 0.0 & 0.0 & 0.0 & 0.0 \\
-100.0 & 0.0 & -5.0 & 0.0 & 0.0 & 0.0 & 0.0 & 0.0 \\
2.0 & 0.0 & 0.0 & 0.0 & -2.0 & 0.0 & 0.0 & 0.0 \\
0.0 & 0.0 & 0.0 & 0.0 & -0.1125 & 0.125 & 0.0 & -0.125 \\
0.0 & 0.0 & 0.0 & 0.0 & 0.0 & -1.67 & 1.67 & 0.0 \\
0.0 & 0.0 & 0.0 & 0.0 & -53.33 & 0.0 & -3.33 & 0.0 \\
2.0 & 0.0 & 0.0 & 0.0 & -2.0 & 0.0 & 0.0 & 0.0
\end{bmatrix}'
$$

$$
B = \begin{bmatrix}
0.0 & 0.0 \\
0.0 & 0.0 \\
5.0 & 0.0 \\
0.0 & 0.0 \\
0.0 & 0.0 \\
0.0 & 0.0 \\
0.0 & 3.33 \\
0.0 & 0.0
\end{bmatrix}, \quad \text{and} \quad F = \begin{bmatrix}
-0.1 & 0.0 \\
0.0 & 0.0 \\
0.0 & 0.0 \\
0.0 & 0.0 \\
0.0 & -0.125 \\
0.0 & 0.0 \\
0.0 & 0.0 \\
0.0 & 0.0
\end{bmatrix}
$$

The initial dynamic output feedback gain matrix $M_0$ is selected as

$$
M_0 = \begin{bmatrix}
-0.1005 & -0.1005 & -0.1005 \\
-0.1005 & 0.1005 & 0.001 \\
-15.0675 & 1.0045 & 0.1196
\end{bmatrix}
$$

and the weighting matrices of the performance index are chosen as $Q_d = 15I$ and $R_d = 10I$, where $I$ is an identity matrix with an appropriate dimension. The matrix of initial conditions $\overline{W}$ is selected as zero except for $\overline{W}(2,2) = \overline{W}(2,1) = 10^5$, and $\gamma = 10^{-4}$. Using the foregoing iterative procedure, the optimal gain matrix is obtained as

$$
M = \begin{bmatrix}
0.0501 & -0.7971 & -0.0966 \\
-0.2323 & 0.8073 & -0.0005 \\
-14.9901 & 0.8224 & 0.1235
\end{bmatrix}
$$

The simulation results of the designed output feedback controller are given in Figure 2.9.

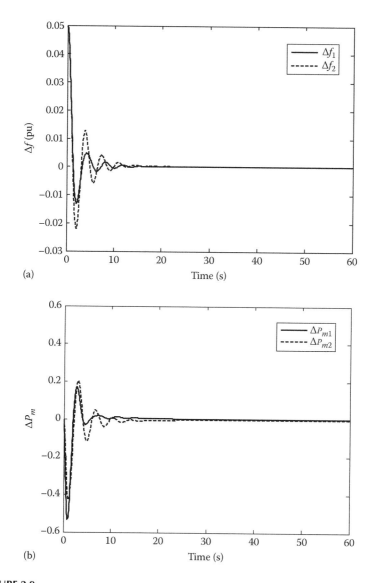

**FIGURE 2.9**
Simulation results of dynamic output feedback for the two-area power system: (a) Frequency deviation and (b) Mechanical power deviation.      *(Continued)*

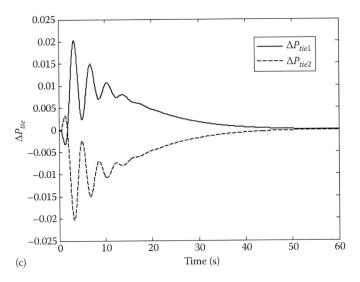

(c)

**FIGURE 2.9 (*Continued*)**
Simulation results of dynamic output feedback for the two-area power system: (c) Tie-line power deviation.

**Example 2.6**

In this example, we design a dynamic output feedback controller for the two-area power system in the previous example, where the outputs of interest are $y_1 = \left[ ACE_1 \ \int ACE_1 \, dt \right]^T$ and $y_2 = \left[ ACE_2 \ \int ACE_2 \, dt \right]^T$ and ACE is the area control error. The model matrices in this case are (see Equation 2.38)

$$
A = \begin{bmatrix}
-0.06 & 0.1 & 0.0 & -0.1 & 0.0 & 0.0 & 0.0 & 0.0 & 0.0 & 0.0 \\
0.0 & -2.0 & 2.0 & 0.0 & 0.0 & 0.0 & 0.0 & 0.0 & 0.0 & 0.0 \\
-100.0 & 0.0 & -5.0 & 0.0 & 0.0 & 0.0 & 0.0 & 0.0 & 0.0 & 0.0 \\
2.0 & 0.0 & 0.0 & 0.0 & 0.0 & -2.0 & 0.0 & 0.0 & 0.0 & 0.0 \\
20.6 & 0.0 & 0.0 & 1.0 & 0.0 & 0.0 & 0.0 & 0.0 & 0.0 & 0.0 \\
0.0 & 0.0 & 0.0 & 0.0 & 0.0 & -0.1125 & 0.125 & 0.0 & -0.125 & 0.0 \\
0.0 & 0.0 & 0.0 & 0.0 & 0.0 & 0.0 & -1.67 & 1.67 & 0.0 & 0.0 \\
0.0 & 0.0 & 0.0 & 0.0 & 0.0 & -53.33 & 0.0 & -3.33 & 0.0 & 0.0 \\
-2.0 & 0.0 & 0.0 & 0.0 & 0.0 & 2.0 & 0.0 & 0.0 & 0.0 & 0.0 \\
0.0 & 0.0 & 0.0 & 0.0 & 0.0 & 16.9 & 0.0 & 0.0 & 1.0 & 0.0
\end{bmatrix},
$$

$$
B = \begin{bmatrix} 0.0 & 0.0 \\ 0.0 & 0.0 \\ 5.0 & 0.0 \\ 0.0 & 0.0 \\ 0.0 & 0.0 \\ 0.0 & 0.0 \\ 0.0 & 0.0 \\ 0.0 & 3.33 \\ 0.0 & 0.0 \\ 0.0 & 0.0 \end{bmatrix}, \quad F = \begin{bmatrix} -0.1 & 0.0 \\ 0.0 & 0.0 \\ 0.0 & 0.0 \\ 0.0 & 0.0 \\ 0.0 & 0.0 \\ 0.0 & -0.125 \\ 0.0 & 0.0 \\ 0.0 & 0.0 \\ 0.0 & 0.0 \\ 0.0 & 0.0 \end{bmatrix}, \quad \text{and}
$$

$$
C = \begin{bmatrix} 20.6 & 0.0 & 0.0 & 1.0 & 0.0 & 0.0 & 0.0 & 0.0 & 0.0 & 0.0 \\ 0.0 & 0.0 & 0.0 & 0.0 & 1.0 & 0.0 & 0.0 & 0.0 & 0.0 & 0.0 \\ 0.0 & 0.0 & 0.0 & 0.0 & 0.0 & 16.9 & 0.0 & 0.0 & 1.0 & 0.0 \\ 0.0 & 0.0 & 0.0 & 0.0 & 0.0 & 0.0 & 0.0 & 0.0 & 0.0 & 1.0 \end{bmatrix}
$$

The matrices $Q_d$ and $R_d$ are selected as $15I$ and $10I$, respectively and the initial condition matrix $\bar{W}$ is selected as zero except for $\bar{W}(2,2) = \bar{W}(2,1) = 10^5$, and $\gamma = 10^{-3}$. The initial feedback gain matrix

$$
M_0 = \begin{bmatrix} -0.15 & 0.3 & 0.30 & 0.05 & -0.05 \\ -0.01 & -0.15 & 0.152 & 0.1 & -0.01 \\ 0.01 & 0.01 & -0.05 & 0.1 & 0.1 \end{bmatrix}
$$

is selected and the final gain matrix

$$
M = \begin{bmatrix} -0.1005 & 0.2675 & 0.3881 & 0.1762 & 0.0614 \\ 0.0128 & -0.1341 & 0.1754 & 0.1297 & -0.0534 \\ 0.0068 & 0.0089 & -0.0581 & 0.0845 & 0.1478 \end{bmatrix}
$$

is obtained after nine iterations. A simulation of the two-area system equipped with the designed dynamic output feedback control is carried out when load disturbances of 0.25 and 0.15 are assumed in area 1 and area 2, respectively. The results of the simulation are shown in Figure 2.10a through e.

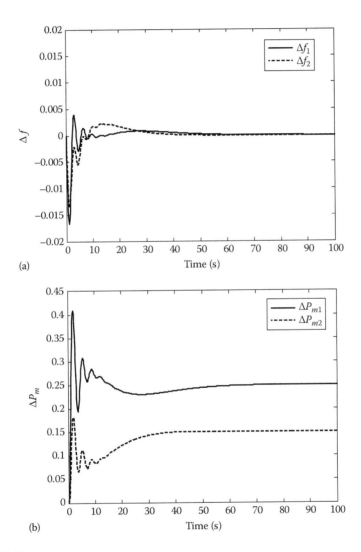

**FIGURE 2.10**
Simulation results of dynamic output feedback for a two-area power system where the output is
the ACE signal: (a) Frequency deviation, (b) Mechanical power deviation.          (*Continued*)

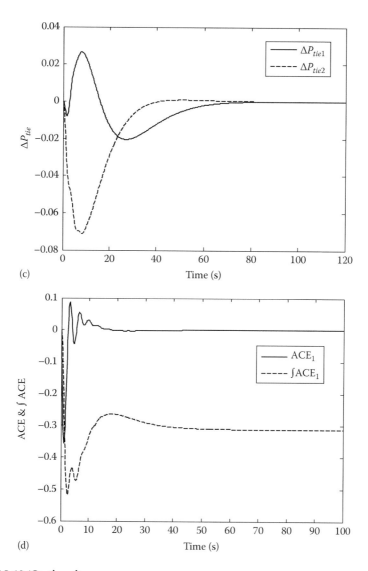

(c)

(d)

**FIGURE 2.10 (*Continued*)**
Simulation results of dynamic output feedback for a two-area power system where the output is
the ACE signal: (c) Tie-line power deviation, (d) ACE and its integration for area 1. (*Continued*)

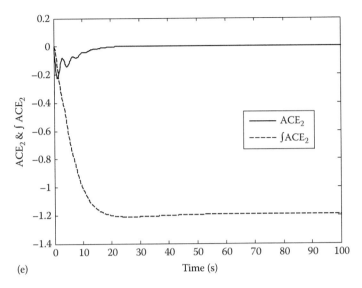

(e)                                                    Time (s)

**FIGURE 2.10 (*Continued*)**
Simulation results of dynamic output feedback for a two-area power system where the output is the ACE signal: (e) ACE and its integration for area 2.

# References

1. R. L. Williams II and D. A. Lawrence, *Linear State-Space Control Systems*, John Wiley & Sons, Inc., Hoboken, NJ, 2007.
2. H. Bevrani, *Robust Power System Frequency Control*, 2nd edn., Springer International Publishing , Cham, Switzerland, 2014.
3. L. Jiang, W. Yao, Q. H. Wu, J. Y. Wen, and S. J. Cheng, Delay-dependent stability for load frequency control with constant and time-varying delays, *IEEE Trans. Power Syst.*, 27(2), 932–941 May 2012.
4. J. Van de Vegte, *Feedback Control Systems*, 3rd edn., Prentice Hall, Upper Saddle River, NJ, 1993.
5. F. L. Lewis, D. L. Vrabie, and V. L. Syrmos, Output feedback and structured control, in *Optimal Control*, 3rd edn., John Wiley & Sons, Inc., Hoboken, NJ, 2012.
6. D. D. Moerder and A. J. Calise, Convergence of a numerical algorithm for calculating optimal feedback gains, *IEEE Trans. Automat. Control*, AC-30(9), 900–903, 1985.

# 3

## LFC of Deregulated Multi-Area Power Systems

### 3.1 Introduction

The power industry worked successfully with a regulated monopoly framework for more than 100 years. The concept of deregulating the power industry framework started in the early 1990s. Deregulation aims to introduce competition at various levels of the power industry. The competitive environment offers a good range of benefits for the customers as well as private entities. Some of the benefits of power industry deregulation would include decreasing electricity price and allowing customers to freely choose between different retailers who struggle to provide a better service. In other words, deregulation or restructuring of the power industry intends to increase competition in wholesale power markets. In this chapter, we present the structure and concept of horizontally integrated utilities. The contract participation and ACE participation factors are introduced to facilitate the formulation of the block diagram model and the deregulated multi-area load frequency control (LFC) power systems.

### 3.2 Deregulated Power System

In the conventional paradigm of power system, the generation, transmission, and distribution are owned by a single entity known as a vertically integrated utility (VIU). In the competitive environment of power systems, the VIU does not exist and a new paradigm is in effect. In this new paradigm, the utilities do not own the generation, transmission, and distribution. Instead, there are three different entities: generation companies (GENCOs), transmission companies (TRANSCOs), and distribution companies (DISCOs). This structure is termed horizontally integrated utility (HIU). As there are several GENCOs and DISCOs in the deregulated structure, any GENCO in any area may supply DISCOs in its user pool and DISCOs in other areas through tie-lines between areas. This means that any DISCO has the liberty to have a contract for transaction of power with any GENCO whether it belongs to its own control area or not. Such transactions are called bilateral transactions and have to be cleared through a neutral entity called an independent system operator (ISO) [1].

The ISO has to control a number of ancillary services pertaining to the quality of the power service. One of the most profitable ancillary services is the automatic generation control (AGC). In a restructured power system, the engineering aspects of planning, operation, and control have to be reformulated although essential ideas remain the same. For example, the common goals of AGC to restore the frequency and the net power interchanges to their desired values remain the same. However, the AGC problem has to be reformulated to take into account the new deregulated HIU structure of the system.

In the deregulated power system, GENCOs sell power to various DISCOs at competitive prices. On the other hand, DISCOs have the freedom to choose the GENCO with the minimum price of power for contracts. Therefore, DISCOs may or may not have contracts with GENCOs in their own area. As a result, various combinations of contracts between GENCOs and DISCOs are possible. The concepts of DISCO participation matrix (DPM) [1–3] and the GENCO participation matrix (GPM) [4–6] have been used to illustrate these bilateral contracts.

### 3.2.1 DISCO Participation Matrix and GENCO Participation Matrix

DPM helps in the formulation of bilateral contracts between DISCOs and GENCOs. The DPM is a matrix having a number of rows equal to the number of GENCOs and a number of columns equal to the number of DISCOs in the system. Each entry of this matrix is a real number less than 1 and is defined as the contract participation factor ($cpf_{kl}$), which is the ratio between the participation of the $k$th GENCO to the total load required by the $l$th DISCO.

Consider a power system consisting of $N$ control areas. The total number of GENCOs and DISCOs in the $i$th area is $n_i$ and $m_i$, respectively. Therefore, the $i$th area consists of $GENCO_{1i}$, $GENCO_{2i}$, ... , $GENCO_{nii}$ and $DISCO_{1i}$, $DISCO_{2i}$, ... , $DISCO_{mii}$, as shown in Figure 3.1. The DPM between area $i$ and area $j$ takes the following form:

$$DPM_{ij} = \begin{bmatrix} cpf_{(r_i+1)(s_j+1)} & cpf_{(r_i+1)(s_j+2)} & \cdots\cdots & cpf_{(r_i+1)(s_j+m_j)} \\ cpf_{(r_i+2)(s_j+1)} & cpf_{(r_i+2)(s_j+2)} & \cdots\cdots & cpf_{(r_i+2)(s_j+m_j)} \\ \vdots & \vdots & \ddots & \vdots \\ cpf_{(r_i+n_i)(s_j+1)} & cpf_{(r_i+n_i)(s_j+2)} & \cdots\cdots & cpf_{(r_i+n_i)(s_j+m_j)} \end{bmatrix} \quad (3.1)$$

where $i,j=1, ... ,N$ and $r_i = \sum_{k=1}^{i-1} n_k$, $s_j = \sum_{k=1}^{j-1} m_k$, and $r_1 = s_1 = 0$.

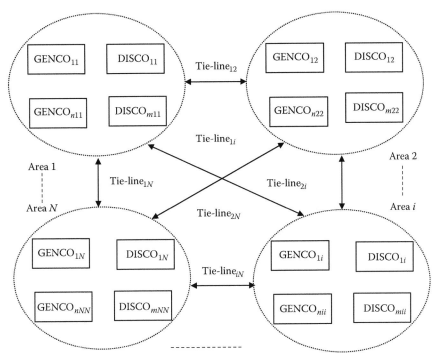

**FIGURE 3.1**
*N*-area deregulated power system.

The augmented distribution participation matrix (ADPM) defined as

$$ADPM = \begin{bmatrix} DPM_{11} & DPM_{12} & \dots\dots\dots & DPM_{1N} \\ DPM_{21} & DPM_{22} & \dots\dots\dots & DPM_{2N} \\ \vdots & \vdots & \ddots & \vdots \\ DPM_{N1} & DPM_{N2} & \dots\dots\dots & DPM_{NN} \end{bmatrix} \tag{3.2}$$

is introduced to consider the participation of all GENCOs of the *N* areas. The diagonal submatrices of ADPM correspond to the participation of local GENCOs to the load demands required by the local DISCOs. The off-diagonal submatrices correspond to the participation of the *i*th GENCO to the load demands required by the *j*th DISCO where $i \neq j$.

**Example 3.1**

As an illustrative example, consider a two-area power system where each area consists of two GENCOs and two DISCOs, as shown in Figure 3.2. Suppose that the load required from each DISCO and the corresponding participation of each GENCO are as shown in Table 3.1.

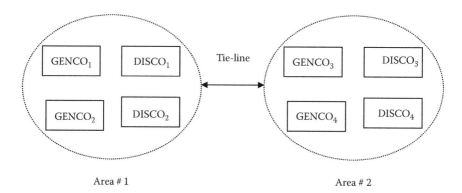

Area # 1                                                              Area # 2

**FIGURE 3.2**
Two-area power system in a deregulated environment.

Find

    a. The DPM of each area
    b. The ADPM of the overall system

**Solution:**
The data shown in Table 3.1 means that $DISCO_1$ has signed contracts with $GENCO_1$ and $GENCO_3$ to supply 0.06 pu and 0.04 pu, respectively. This means that the contract participation factors $cpf_{11}=0.06/0.1=0.6$ and $cpf_{31}=0.04/0.1=0.4$. The other rows of Table 3.1 can be explained in a similar way.

    a. The DPM of each area is calculated as

$$DPM_{11} = \begin{bmatrix} cpf_{11} & cpf_{12} \\ cpf_{21} & cpf_{22} \end{bmatrix} = \begin{bmatrix} \dfrac{0.06}{0.1} & 0.0 \\ 0.0 & \dfrac{0.15}{0.25} \end{bmatrix} = \begin{bmatrix} 0.6 & 0.0 \\ 0.0 & 0.6 \end{bmatrix} \tag{3.3}$$

$$DPM_{12} = \begin{bmatrix} cpf_{13} & cpf_{14} \\ cpf_{23} & cpf_{24} \end{bmatrix} = \begin{bmatrix} 0.0 & \dfrac{0.1}{0.3} \\ 0.0 & \dfrac{0.15}{0.3} \end{bmatrix} = \begin{bmatrix} 0.0 & 0.33 \\ 0.0 & 0.5 \end{bmatrix} \tag{3.4}$$

**TABLE 3.1**

Participation of Different GENCOs to Satisfy Demands Required by Different DISCOs

|  | GENCO$_1$ | GENCO$_2$ | GENCO$_3$ | GENCO$_4$ |
|---|---|---|---|---|
| DISCO$_1$ = 0.1 | 0.06 | 0.0 | 0.04 | 0.0 |
| DISCO$_2$ = 0.25 | 0.0 | 0.15 | 0.05 | 0.05 |
| DISCO$_3$ = 0.05 | 0.0 | 0.0 | 0.05 | 0.0 |
| DISCO$_4$ = 0.3 | 0.1 | 0.15 | 0.0 | 0.05 |

$$DPM_{21} = \begin{bmatrix} cpf_{31} & cpf_{32} \\ cpf_{41} & cpf_{42} \end{bmatrix} = \begin{bmatrix} 0.04 & 0.05 \\ \dfrac{}{0.1} & 0.25 \\ 0.0 & \dfrac{0.05}{0.25} \end{bmatrix} = \begin{bmatrix} 0.4 & 0.2 \\ 0.0 & 0.2 \end{bmatrix} \qquad (3.5)$$

$$DPM_{22} = \begin{bmatrix} cpf_{33} & cpf_{34} \\ cpf_{43} & cpf_{44} \end{bmatrix} = \begin{bmatrix} \dfrac{0.05}{0.05} & 0.0 \\ 0.0 & \dfrac{0.05}{0.3} \end{bmatrix} = \begin{bmatrix} 1 & 0.0 \\ 0.0 & 0.17 \end{bmatrix} \qquad (3.6)$$

b. For $N=2$, the ADPM is given by

$$ADPM = \begin{bmatrix} DPM_{11} & DPM_{12} \\ DPM_{21} & DPM_{22} \end{bmatrix} \qquad (3.7)$$

Substituting (3.3) through (3.6) into (3.7), we get

$$ADPM = \begin{bmatrix} cpf_{11} & cpf_{12} & cpf_{13} & cpf_{14} \\ cpf_{21} & cpf_{22} & cpf_{23} & cpf_{24} \\ cpf_{31} & cpf_{32} & cpf_{33} & cpf_{34} \\ cpf_{41} & cpf_{42} & cpf_{43} & cpf_{44} \end{bmatrix} = \begin{bmatrix} 0.6 & 0.0 & 0.0 & 0.33 \\ 0.0 & 0.6 & 0.0 & 0.5 \\ 0.4 & 0.2 & 1.0 & 0.0 \\ 0.0 & 0.2 & 0.0 & 0.17 \end{bmatrix} \qquad (3.8)$$

Note that the sum of the column entries of ADPM is unity, that is, $\sum_{k=1}^{4} cpf_{kl} = 1$. The diagonal submatrix $DPM_{11}$ of ADPM corresponds to the participation of $GENCO_1$ and $GENCO_2$ of area 1 to the load demands required by the local $DISCO_1$ and $DISCO_2$ and likewise the diagonal submatrix $DPM_{22}$ corresponds to the participation of $GENCO_3$ and $GENCO_4$ of area 2 to the load demands required by the local $DISCO_3$ and $DISCO_4$. The off-diagonal submatrix $DPM_{12}$ ($DPM_{21}$) corresponds to the participation of GENCOs in area 1 (area 2) to the load demands required by DISCOs in area 2 (area 1). In general, for N-area power systems, the diagonal submatrices of the ADPM represent the contracted contribution of local GENCOs to the load requirement of the local DISCOs. The off-diagonal submatrices of ADPM correspond to the contracted participation of GENCOs of a specific area to the load demand of DISCOs in other areas.

In the literature, there is another participation matrix used to represent the contracted power between GENCOs and DISCOs. This matrix is termed as the augmented GENCOs participation matrix (AGPM). It has the same structure as the ADPM [5]. Each entry of the GENCOs participation matrix is defined as the generation participation factor ($gpf_{ij}$). The $gpf_{ij}$ represents the participation of $GENCO_i$ to the total load of $DISCO_j$. Since both $cpf_{ij}$ and $gpf_{ij}$ have effectively the same meaning, the use of either DPM or GPM to visualize the bilateral contracts between DISCOs and GENCOs is equivalent. In reality, the DISCOs sign contracts with GENCOs to buy power at the most competitive price available; therefore, the use of DPM is more expressive than the GPM.

## 3.3  Block Diagram Model of a Deregulated Multi-Area LFC System

In this section, we develop a block diagram for multi-area LFC in a deregulated environment. Consider an N-area conventional power system where each area is assumed to have $n_i$ generator units. As there are more than one generator in each area, the ACE signal for each area, $ACE_i, i=1, \dots, N$, has to be distributed among these generators in proportion to their participation in the AGC. The coefficients that describe the participation of each generator to the $ACE_i$ signal are termed as the ACE participation factor (*apfs*). The sum of all *apfs* in a given area is unity, that is, $\sum_{j=1}^{n_i} apf_j = 1$. In this case, the total $ACE_i$ is written as

$$ACE_i = \Delta P_{tiei} + B_i \Delta f_i \tag{3.9}$$

where $\Delta P_{tiei}$ is defined as the total tie-line power deviation from the *i*th area and is given by

$$\Delta P_{tiei} = \frac{2\pi}{s} \left( \sum_{\substack{j=1 \\ j \neq i}}^{N} T_{ij} \Delta f_i - \sum_{\substack{j=1 \\ j \neq i}}^{N} T_{ij} \Delta f_j \right) \tag{3.10}$$

where
  $B_i$ is the bias factor
  $\Delta f_i$ is the frequency deviation

According to the *apfs* of each generator, the speed changer setting for the *k*th generator in the *i*th area $\Delta P_{cki}$ takes the form

$$\Delta P_{cki} = \left( apf_{k_i} \right) ACE_i \tag{3.11}$$

The complete block diagram for the conventional LFC system of the *i*th area is shown in Figure 3.3, where $\Delta P_{di}$ represents the load disturbance.

Now, consider the *i*th area of the *N*-area deregulated power system shown in Figure 3.1. The load change required by a DISCO is represented by a local load demand $\Delta P_{Loci}$ in the area to which this DISCO belongs. The total load demand of the *i*th area $d_i$ is formulated as the sum of the total contracted demand $\Delta P_{Loci}$ and the total uncontracted demand $\Delta P_{di}$ [4], that is,

$$d_i = \Delta P_{Loci} + \Delta P_{di} \tag{3.12}$$

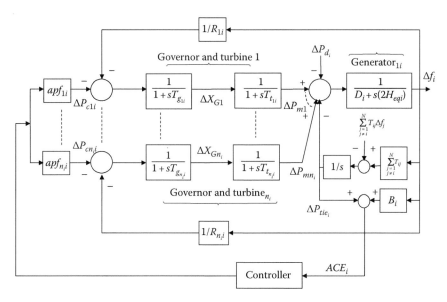

**FIGURE 3.3**
LFC scheme of area $i$ in a conventional power system.

with

$$\Delta P_{Loci} = \sum_{j=1}^{m_i} \Delta P_{Lji} \qquad (3.13)$$

$$\Delta P_{di} = \sum_{j=1}^{m_i} \Delta P_{ULji} \qquad (3.14)$$

where $\Delta P_{Lji}$ and $\Delta P_{ULji}$ represent the contracted and uncontracted demands, respectively, for the $j$th DISCO in the $i$th area.

The scheduled power $\Delta P_{tie\ scheduled_{ij}}$ flowing in the tie-line connecting area $i$ and area $j$ is formulated as the difference between the total demand of DISCOs in area $i$ from GENCOs in area $j$ and the total demand of DISCOs in area $j$ from GENCOs in area $i$. Therefore, $\Delta P_{tie\ scheduled_{ij}}$ can be written as

$$\Delta P_{tie\ scheduled_{ij}} = \sum_{q=1}^{n_i}\sum_{r=1}^{m_j} cpf_{(s_i+q)(z_j+r)}\Delta P_{Lrj} - \sum_{r=1}^{n_j}\sum_{q=1}^{m_i} cpf_{(s_j+r)(z_i+j)}\Delta P_{Lji} \qquad (3.15)$$

with $z_i = \sum_{k=1}^{i-1} n_k$, $s_j = \sum_{k=1}^{j-1} m_k$ and $z_1 = s_1 = 0$. To illustrate (3.15), consider the two-area system shown in Figure 3.2. In this case $n_1 = n_2 = 2$, $m_1 = m_2 = 2$, and (3.15) becomes

$$\Delta P_{tie\,scheduled_{ij}} = \sum_{q=1}^{2}\sum_{r=1}^{2} cpf_{(s_i+q)(z_j+r)}\Delta P_{Lrj} - \sum_{r=1}^{2}\sum_{q=1}^{2} cpf_{(s_j+r)(z_i+q)}\Delta P_{Ljq}$$

$$= \left( cpf_{11}\Delta P_{L12} + cpf_{12}\Delta P_{L22} + cpf_{21}\Delta P_{L12} + cpf_{22}\Delta P_{L22} \right)$$

$$- \left( cpf_{11}\Delta P_{L21} + cpf_{12}\Delta P_{L22} + cpf_{21}\Delta P_{L21} + cpf_{22}\Delta P_{L22} \right)$$

For the *N*-area system, the total scheduled tie-line power $\ell_i$ flowing from other areas to the *i*th area is defined as

$$\ell_i = \sum_{j=1, j\neq i}^{N} \Delta P_{tie\,scheduled_{ij}} \tag{3.16}$$

Moreover, the tie-line power error $\Delta P_{tie\,error_i}$ is defined as

$$\Delta P_{tie\,error_i} = \Delta P_{tie_i} - \ell_i \tag{3.17}$$

where $\Delta P_{tie_i}$ is the actual tie-line power flow. This is different from the conventional LFC scheme where there is no scheduled tie-line power flow $\ell_i$. Hence, in the deregulated paradigm, the contracts between different GENCOs and DISCOs affect not only the load demand of each area but also the exchange tie-line power flow. In this case, the area control error for the *i*th area $ACE_i$ is given by

$$ACE_i = \Delta P_{tie\,error_i} + B_i\Delta f_i \tag{3.18}$$

Combining (3.17) and (3.18), the following $ACE_i$ equation is obtained:

$$ACE_i = \Delta P_{tie_i} - \ell_i + B_i\Delta f_i \tag{3.19}$$

As there are more than one GENCO in each area, the $ACE_i$ signal has to be distributed among them in proportion to their participation in the AGC. The ACE participation factor in the deregulated environment describes the participation of each GENCO in area *i* to the overall $ACE_i$ signal. In the steady state, generation of a GENCO must match the contracted demand of DISCOs. The contracted generation in pu MW can be expressed in matrix form using the *ADPM* as

$$\Delta P_{Gc} = (ADPM)\Delta P_{Lc} \tag{3.20}$$

where $\Delta P_{Gc}$ and $\Delta P_{Lc}$ are the matrices of the generation of each GENCO and the load demand of each DISCO, respectively. A block diagram of the *i*th area

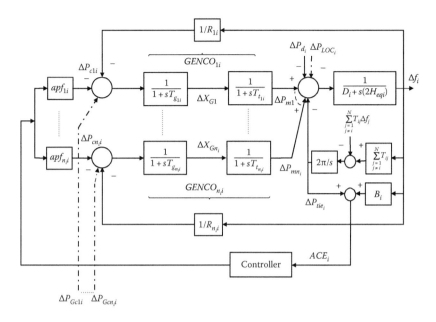

**FIGURE 3.4**
LFC scheme of area *i* in a deregulated LFC environment.

in deregulated scenario is shown in Figure 3.4. In this figure, the signal $\Delta P_{Gcki}$, $k=1,\dots,n_i$ represents one row of the matrix (3.20), which can be written as

$$\Delta P_{Gcki} = \sum_{j=1}^{N}\sum_{h=1}^{m_j} cpf_{(r_i+k)(s_j+h)}\Delta P_{Lckj} \tag{3.21}$$

**Example 3.2**

Consider a three-area deregulated power system having the data shown in Table 3.2 with the participation of different GENCOs to supply the DISCO demands given in Table 3.3.

1. Determine the ADPM
2. Find the steady-state generation of each GENCO to satisfy the DISCO load demands.

**Solution:**

1. Since the number of areas $N = 3$, using (3.2), the ADPM will be

$$ADPM = \begin{bmatrix} DPM_{11} & DPM_{12} & DPM_{13} \\ DPM_{21} & DPM_{22} & DPM_{23} \\ DPM_{31} & DPM_{32} & DPM_{33} \end{bmatrix}$$

**TABLE 3.2**

Data of a Three-Area Deregulated Power System

| Area | Number of GENCOs ($n_i$) | Number of DISCOs ($m_i$) | Load Demand (pu) in Each DISCO |
|---|---|---|---|
| 1 | 3 | 1 | $\Delta P_{L1} = 0.2$ |
| 2 | 2 | 2 | $\Delta P_{L2_1} = 0.1$ and $\Delta P_{L2_2} = 0.25$ |
| 3 | 1 | 3 | $\Delta P_{L3_1} = 0.1, \Delta P_{L3_2} = 0.075$ and $\Delta P_{L3_3} = 0.05$ |

**TABLE 3.3**

Participation of Different GENCOs to Satisfy Demands Required by Different DISCOs for a Three-Area System

| GENCOs → | $\Delta P_{Gc1_1}$ | $\Delta P_{Gc2_1}$ | $\Delta P_{Gc3_1}$ | $\Delta P_{Gc1_2}$ | $\Delta P_{Gc2_2}$ | $\Delta P_{Gc1_3}$ |
|---|---|---|---|---|---|---|
| DISCOs↓ | | | | | | |
| $\Delta P_{Lc1_1} = 0.1$ | 0.06 | 0.0 | 0.015 | 0.0 | 0.025 | 0.0 |
| $\Delta P_{Lc1_2} = 0.25$ | 0.025 | 0.1 | 0.05 | 0.05 | 0.0 | 0.025 |
| $\Delta P_{Lc2_2} = 0.15$ | 0.0 | 0.0 | 0.0 | 0.05 | 0.0 | 0.1 |
| $\Delta P_{Lc1_3} = 0.35$ | 0.0 | 0.0 | 0.0 | 0.0 | 0.1 | 0.25 |
| $\Delta P_{Lc2_3} = 0.2$ | 0.075 | 0.025 | 0.0 | 0.0 | 0.0 | 0.1 |
| $\Delta P_{Lc3_3} = 0.05$ | 0.0 | 0.0 | 0.0 | 0.0 | 0.0 | 0.05 |

where the submatrices $DPM_{ij}$, $i=1,2,3$ and $j=1,2,3$ are determined using (3.1) and the data of Table 3.3 as follows:

$$DPM_{11} = \begin{bmatrix} 0.06/0.1 \\ 0.0 \\ 0.015/0.1 \end{bmatrix} = \begin{bmatrix} 0.6 \\ 0.0 \\ 0.15 \end{bmatrix},$$

$$DPM_{12} = \begin{bmatrix} 0.025/0.25 & 0.0 \\ 0.1/0.25 & 0.0 \\ 0.05/0.25 & 0.0 \end{bmatrix} = \begin{bmatrix} 0.1 & 0.0 \\ 0.4 & 0.0 \\ 0.2 & 0.0 \end{bmatrix},$$

$$DPM_{13} = \begin{bmatrix} 0.0 & 0.075/0.2 & 0.0 \\ 0.0 & 0.025/0.2 & 0.0 \\ 0.0 & 0.0 & 0.0 \end{bmatrix} = \begin{bmatrix} 0.0 & 0.375 & 0.0 \\ 0.0 & 0.125 & 0.0 \\ 0.0 & 0.0 & 0.0 \end{bmatrix},$$

$$DPM_{21} = \begin{bmatrix} 0.0 \\ 0.025/0.1 \end{bmatrix} = \begin{bmatrix} 0.0 \\ 0.25 \end{bmatrix},$$

$$DPM_{22} = \begin{bmatrix} 0.05/0.25 & 0.05/0.15 \\ 0.0 & 0.0 \end{bmatrix} = \begin{bmatrix} 0.2 & 0.33 \\ 0.0 & 0.0 \end{bmatrix},$$

$$DPM_{23} = \begin{bmatrix} 0.0 & 0.0 & 0.0 \\ 0.1/0.35 & 0.0 & 0.0 \end{bmatrix},$$

$$DPM_{31} = \begin{bmatrix} 0.0 \end{bmatrix}, \quad DPM_{32} = \begin{bmatrix} 0.025/0.25 & 0.1/0.15 \end{bmatrix} = \begin{bmatrix} 0.1 & 0.67 \end{bmatrix}, \quad \text{and}$$

$$DPM_{33} = \begin{bmatrix} 0.25/0.35 & 0.1/0.2 & 0.05/0.05 \end{bmatrix} = \begin{bmatrix} 0.714 & 0.5 & 1.0 \end{bmatrix}.$$

Therefore,

$$ADPM = \begin{bmatrix} 0.6 & 0.1 & 0.0 & 0.0 & 0.375 & 0.0 \\ 0.0 & 0.4 & 0.0 & 0.0 & 0.125 & 0.0 \\ 0.15 & 0.2 & 0.0 & 0.0 & 0.0 & 0.0 \\ 0.0 & 0.2 & 0.33 & 0.0 & 0.0 & 0.0 \\ 0.25 & 0.0 & 0.0 & 0.286 & 0.0 & 0.0 \\ 0.0 & 0.1 & 0.67 & 0.714 & 0.5 & 1.0 \end{bmatrix}.$$

Note that the sum of entries of each column in the ADPM is unity.

2. Using (3.20), we find the contracted generation as follows:

$$\begin{bmatrix} GENCOs\ area1 \begin{cases} \Delta P_{Gc1_1} \\ \Delta P_{Gc2_1} \\ \Delta P_{Gc3_1} \end{cases} \\ \cdots\cdots\cdots\cdots\cdots\cdots \\ GENCOs\ area2 \begin{cases} \Delta P_{Gc1_2} \\ \Delta P_{Gc2_2} \end{cases} \\ \cdots\cdots\cdots\cdots\cdots\cdots \\ GENCOs\ area3 \{ \Delta P_{Gc1_3} \end{bmatrix} = \begin{bmatrix} DPM_{11} & DPM_{12} & DPM_{13} \\ DPM_{21} & DPM_{22} & DPM_{23} \\ DPM_{31} & DPM_{32} & DPM_{33} \end{bmatrix} \begin{bmatrix} DISCOs\ area1 \{ \Delta P_{Lc1_1} \\ \cdots\cdots\cdots\cdots\cdots\cdots \\ DISCOs\ area2 \begin{cases} \Delta P_{Lc1_2} \\ \Delta P_{Lc2_2} \end{cases} \\ \cdots\cdots\cdots\cdots\cdots\cdots \\ DISCOs\ area3 \begin{cases} \Delta P_{Lc1_3} \\ \Delta P_{Lc2_3} \\ \Delta P_{Lc3_3} \end{cases} \end{bmatrix}$$

Using the obtained ADPM, the steady state generation of each GENCO to satisfy the contracted DISCO load demands are determined as

$$\begin{bmatrix} \Delta P_{Gc1_1} \\ \Delta P_{Gc2_1} \\ \Delta P_{Gc3_1} \\ \Delta P_{Gc1_2} \\ \Delta P_{Gc2_2} \\ \Delta P_{Gc1_3} \end{bmatrix} = \begin{bmatrix} 0.6 & 0.1 & 0.0 & 0.0 & 0.375 & 0.0 \\ 0.0 & 0.4 & 0.0 & 0.0 & 0.125 & 0.0 \\ 0.15 & 0.2 & 0.0 & 0.0 & 0.0 & 0.0 \\ 0.0 & 0.2 & 0.333 & 0.0 & 0.0 & 0.0 \\ 0.25 & 0.0 & 0.0 & 0.286 & 0.0 & 0.0 \\ 0.0 & 0.1 & 0.666 & 0.714 & 0.5 & 0.0 \end{bmatrix} \begin{bmatrix} 0.1 \\ 0.25 \\ 0.15 \\ 0.35 \\ 0.2 \\ 0.05 \end{bmatrix} = \begin{bmatrix} 0.16 \\ 0.125 \\ 0.065 \\ 0.0995 \\ 0.1251 \\ 0.4754 \end{bmatrix}$$

## References

1. V. Donde, M. A. Pai, and I. A. Hiskens, Simulation and optimization in an AGC system after deregulation, *IEEE Trans. Power Syst.*, 16(3), 481–489, August 2001.
2. J. Kumar, K. Ng, and G. Sheble, AGC simulator for price-based operation: Part I, *IEEE Trans. Power Syst.*, 12(2), 527–532, May 1997.
3. J. Kumar, K. Ng, and G. Sheble, AGC simulator for price-based operation: Part II, *IEEE Trans. Power Syst.*, 12(2), 533–538, May 1997.
4. H. Bevrani, Y. Mitani, and K. Tsuji, Robust decentralized AGC in a restructured power system, *Energy Convers. Manage.*, 45, 2297–2312, 2004.
5. W. Tan, H. Zhang, and M. Yu, Decentralized load frequency control in deregulated environments, *Electr. Power Energy Syst.*, 41, 16–26, 2012.
6. H. Shayeghi, H. A. Shayanfar, and O. P. Malik, Robust decentralized neural networks based LFC in a deregulated power system, *Electr. Power Syst. Res.*, 77, 241–251, 2007.

## Further Readings

1. E. De Tuglie and F. Torelli, Load following control schemes for deregulated energy markets, *IEEE Trans. Power Syst.*, 21(4), 1691–1698, November 2006.
2. I. P. Kumar and D. P. Kothari, Recent philosophies of automatic generation control strategies in power systems, *IEEE Trans. Power Syst.*, 20(1), 346–357, February 2005.
3. B. H. Bakken and O. S. Grande, Automatic generation control in a deregulated power system, *IEEE Trans. Power Syst.*, 13(4), 1401–1406, November 1998.
4. E. Rakhshani and J. Sadeh, Practical viewpoints on load frequency control problem in a deregulated power system, *Energy Convers. Manage.*, 51, 1148–1156, 2010.
5. H. Shayeghi, A. Jalili, and H. A. Shayanfar, A robust mixed H2/H1 based LFC of a deregulated power system including SMES, *Energy Convers. Manage.*, 49, 2656–2668, 2008.
6. W. Tan, Y. Hao, and D. Li, Load frequency control in deregulated environments via active disturbance rejection, *Electr. Power Energy Syst.*, 66, 166–177, 2015.

# 4

## PID LFC Controllers

## 4.1 Introduction

The proportional–integral–derivative (PID) control offers the simplest and yet most efficient solution to any real-world control problems. The three terms of the PID controller provide improvement to both the transient and steady-state specifications of the control system response. Due to their simplicity and functionality, PID controllers became very popular since the development of the Ziegler–Nichols' (Z-N) tuning method in 1942 [1]. In this chapter, we present an introduction about the structure, design, and tuning of PID controllers. A model-based PID design via the internal model control is explained. Then, we move on to the application of these controllers to the LFC problem of single area. Finally, model reduction techniques are presented to facilitate the PID controller design.

## 4.2 PID Controller Architecture

A block diagram of a plant with a PID controller is shown in Figure 4.1. The PID controller has three principal control effects. The proportional term ($P$) corresponds to the change of the control effort ($u$) proportional to the error signal ($e$). The integral term ($I$) gives a control action that is proportional to the time integral of the error. The integral term ensures the elimination of the steady-state error. The derivative term ($D$) provides a control signal proportional to the time derivative of the error signal. This term allows prediction of errors, is commonly used to increase the system damping, and hence has a stabilizing effect.

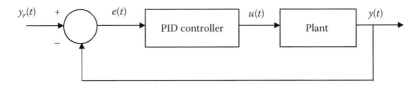

**FIGURE 4.1**
PID controller and plant in unity feedback configuration.

A practical difficulty with PID control design is the lack of industrial standards. As a result, there are many variations of the basic PID algorithm that will substantially improve its tuning, performance, and operation. There are a wide variety of PID controller architectures in the literature. Seven different structures for the PI controller and forty-six different structures for the PID controller have been identified [2]. The common architectures of a PID controller [3,4] are explained next.

## 4.2.1 The Ideal PID Controller

This controller is described by [3]

$$u(t) = K_c \left( e(t) + \frac{1}{T_i} \int_0^t e(\tau) d\tau + T_d \frac{de(t)}{dt} \right) \tag{4.1}$$

or, in the transfer function form

$$G_i(s) = \frac{U(s)}{E(s)} = K_c \left( 1 + \frac{1}{T_i s} + T_d s \right) \tag{4.2}$$

where $K_c$, $T_i$, and $T_d$ are the controller gain, the integral time, and the derivative time, respectively. A block diagram of a PID controller is shown in Figure 4.2.

This controller structure is also termed as the standard, parallel, non-interacting or ISA (Instrumentation, Systems, and Automation Society) controller [2,3].

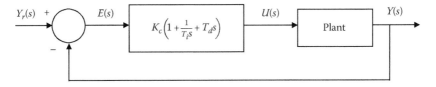

**FIGURE 4.2**
Ideal PID controller in a unity feedback block diagram representation.

## 4.2.2 The Classical PID Controller

The transfer function of this controller is given as

$$G_s(s) = \frac{U(s)}{E(s)} = K_{c1}\left(1 + \frac{1}{T_{i1}s}\right)(1 + T_{d1}s)$$ (4.3)

where $K_{c1}$, $T_{i1}$, and $T_{d1}$ are the controller gain, the integral time, and the derivative time, respectively. The block diagram representation is shown in Figure 4.3.

This type of controller is also termed as the cascade, interacting, or series PID controller [3]. The controller (4.3) can always be represented as a noninteracting controller (4.2), whose coefficients are given by

$$K_c = K_{c1}\frac{T_{i1} + T_{d1}}{T_{i1}}$$ (4.4)

$$T_i = T_{i1} + T_{d1}$$ (4.5)

$$T_d = \frac{T_{i1}T_{d1}}{T_{i1} + T_{d1}}$$ (4.6)

Equations 4.4 through 4.6 are obtained by comparing the coefficients of $s$ in (4.2) and (4.3). Conversely, the noninteracting PID controller (4.2) can be written in the equivalent interacting controller form (4.3) with the following equivalence:

$$K_{c1} = \frac{K_c}{2}\left(1 + \sqrt{1 - 4\frac{T_d}{T_i}}\right)$$ (4.7)

$$T_{i1} = \frac{T_i}{2}\left(1 + \sqrt{1 - 4\frac{T_d}{T_i}}\right)$$ (4.8)

$$T_{d1} = \frac{T_i}{2}\left(1 - \sqrt{1 - 4\frac{T_d}{T_i}}\right)$$ (4.9)

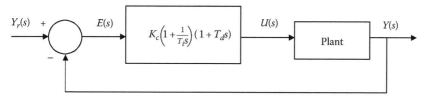

**FIGURE 4.3**
Classical PID controller in a unity feedback block diagram representation.

Investigating Equations 4.7 through 4.9 shows that converting the ideal PID controller to an equivalent classical controller is not always possible unless $T_i \geq 4T_d$.

### 4.2.3 The Parallel PID Controller

In this structure, the three control actions are completely separated and the transfer function is given as

$$G_p(s) = \frac{U(s)}{E(s)} = K_p + \frac{K_i}{s} + K_d s \tag{4.10}$$

In fact, this parallel structure of PID controller is the most general form. This is because the integral action (derivative action) can be switched off by letting $K_i = 0$ ($K_d = 0$). Note that in the ideal and classical structures of PID, to switch off the integral action, the integral time must reach infinity. It is straightforward to convert the ideal controller parameters to the parallel form using the following

$$K_p = K_c \tag{4.11}$$

$$K_i = \frac{K_c}{T_i} \tag{4.12}$$

$$K_d = K_c T_c \tag{4.13}$$

---

## 4.3 Modified PID Controller Architectures

The basic PID controller structures given by (4.2), (4.3), and (4.10) are not suitable for practical applications. There are two main coupled reasons behind this fact. First, the controller transfer function is not proper and therefore it cannot be implemented in practice. This problem is due to the derivative action, which may cause difficulties if there is a high-frequency measurement noise. Consider, for example, a high-frequency sinusoidal noise signal $n(t) = A \sin(\omega t)$ superimposed on the error signal. In the presence of the derivative action, the control signal includes, due to this noise, a term $u_{dn}(t) = AK_c T_d \omega \cos(\omega t)$. Practically speaking, this term of the control signal with high amplitude and frequency may cause damage to the actuator. These problems can be rectified by including at least a first-order low-pass filter to filter out the derivative control action. In this regard, we may have two possibilities. One possibility is to replace the derivative action $T_d s$ of the ideal structure by $T_d s/(1+(T_d/N)s)$ or the term $(1+T_{d1}s)$ of the classical structure by $(1+T_{d1}s)/(1+(T_{d1}/N_1)s)$. Therefore, the modified versions of (4.2) and (4.3) take the following forms, respectively:

$$G_{im1}(s) = K_c\left(1+\frac{1}{T_i s}+\frac{T_d s}{1+(T_d/N)s}\right) \tag{4.14}$$

$$G_{cm}(s) = K_{c1}\left(1+\frac{1}{T_{i1}s}\right)\left(\frac{1+T_{d1}s}{1+(T_{d1}/N_1)s}\right) \tag{4.15}$$

The block diagrams of (4.14) and (4.15) are presented in Figures 4.4 and 4.5, respectively.

The other possibility to overcome the problems associated with the derivative term is to filter out the overall control signal. This can be done by

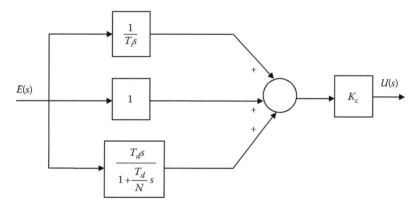

**FIGURE 4.4**
Modified ideal (noninteracting) PID controller.

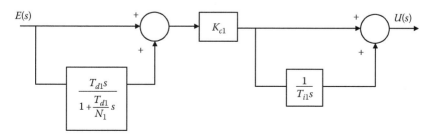

**FIGURE 4.5**
Modified classical (interacting) PID controller.

cascading the PID controller with a first- or a second-order filter [5]. In this context, the modified ideal PID control law (4.2) becomes

$$G_{im2}(s) = K_c \left(1 + \frac{1}{T_i s} + T_d s\right) \frac{1}{(1 + T_f s)} \qquad (4.16)$$

$$G_{im3}(s) = K_c \left(1 + \frac{1}{T_i s} + T_d s\right) \frac{1}{(1 + T_f s)^2} \qquad (4.17)$$

A block diagram of these modified versions is shown in Figure 4.6.

Before concluding this part, it is worth highlighting the following comments:

1. The interacting forms of PID controllers given in (4.3) and (4.15) are mostly used in practice since they can be easily tuned [3].
2. The value $N$ and $N_1$ in (4.14) and (4.15) range between 1 and 33. However, in the majority of applications, the range lies between 8 and 16 [6].
3. If the reference signal $y_r(t)$ experiences a stepwise variation, the derivative action yields a large spike in the control signal. Such large spikes may damage the actuators. This phenomenon is known as a derivative kick. To resolve this issue, the derivative term is applied to the plant output and not to the error signal, as shown in Figure 4.7.

For example, when the derivative part of the classical PID controller with filter $D = (1 + T_{d1}s)/(1 + (T_{d1}/N_1)s)$ is applied to the output, the control signal becomes

$$U(s) = I(Y_r(s) - DY(s)) \qquad (4.18)$$

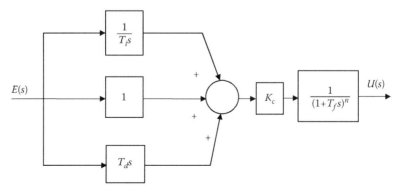

**FIGURE 4.6**
Modified ideal PID controller, $n = 1, 2$.

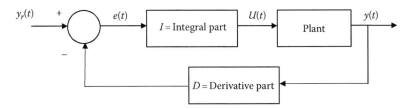

**FIGURE 4.7**
Classical PID with the derivation action applied to the output signal.

where

$$I = K_{c1} \left( 1 + \frac{1}{T_{i1}s} \right)$$

## 4.4 Tuning of PID Controllers

Tuning of PID controllers is a critical issue in control system design. In the PID controller design, several tasks have to be considered. The two main tasks that should be focused on are either the reference-signal tracking or the disturbance rejection. However, in some cases, both of these tasks have to be considered simultaneously. Also, the controller should yield an acceptable level of control signal in order not to harm the actuator. Moreover, the designed controller should perform well in the face of plant parameter uncertainties, that is, the controller should be robust.

There are two main tuning techniques to design the PID controller parameters. These techniques are the Ziegler–Nichols method and an analytical method based on the internal model control principle. In what follows, we give a brief description of each technique.

### 4.4.1 Ziegler–Nichols Tuning Techniques

The most famous tuning methods for PID controllers are those developed by Ziegler and Nichols [1]. They suggested two methods to tune the parameters of PID controllers. These methods are developed based on some information extracted from the plant response.

#### 4.4.1.1 The First Method of Ziegler–Nichols

In this method, the open-loop unit-step response (the response of the plant only) is obtained experimentally or from simulation, as shown in Figure 4.8. Such a response, termed as S-shape response, results from a plant that has no integrator or dominant complex conjugate poles. This response is

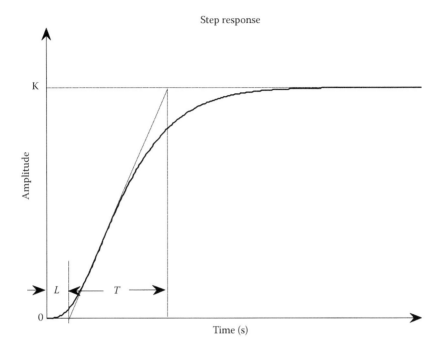

**FIGURE 4.8**
Plant unit step response.

characterized by two parameters: the delay time $L$ and the time constant $T$. In this case, the transfer function of the plant can be approximated by a first-order plus time delay (FOPTD) model as

$$G(s) = \frac{Y(s)}{U(s)} = \frac{Ke^{-Ls}}{1+Ts} \tag{4.19}$$

The parameters of the ideal PID controller, $K_c$, $T_i$, and $T_d$, given in (4.2) are suggested by Ziegler and Nichols in terms of $L$ and $T$, according to Table 4.1.

**TABLE 4.1**

Ziegler–Nichols Tuning Rules Based on Plant Step Response

| Type of Controller | $K_c$ | $T_i$ | $T_d$ |
|---|---|---|---|
| P | $\dfrac{T}{L}$ | $\infty$ | 0 |
| PI | $0.9\dfrac{T}{L}$ | $\dfrac{L}{0.3}$ | 0 |
| PID | $1.2\dfrac{T}{L}$ | $2L$ | $0.5L$ |

### 4.4.1.2 The Second Method of Ziegler–Nichols

In this method, the closed-loop step response is obtained while only the proportional action $K_c$ is activated and the integral time and derivative time are set at $T_i = \infty$ and $T_d = 0$, respectively. The gain $K_c$ is then increased slowly until the system output exhibits sustained oscillations. In this case, the ultimate gain $K_{cu}$ and the ultimate period of oscillations $T_u$ are recorded. Ziegler and Nichols suggested setting the parameters of the ideal-type PID controller $K_c$, $T_i$, and $T_d$ in terms of $K_{cu}$ and $T_u$, according to Table 4.2.

#### Example 4.1

Consider the isolated power area shown in Figure 4.9 with $H_{eq} = 6$, $D = 0.5$, $T_g = 0.2$, $T_t = 0.3$ and $R = 0.05$. Design a PID controller using

1. The first method of Ziegler–Nichols
2. The second method of Ziegler–Nichols
3. Simulate the closed-loop response of each controller when the load disturbance is 20%.

#### Solution:

1. *Design using the first method of Ziegler–Nichols*

**TABLE 4.2**

Ziegler–Nichols Tuning Rules Based on Ultimate Gain and Ultimate Period

| Type of Controller | $K_c$ | $T_i$ | $T_d$ |
| --- | --- | --- | --- |
| P | $0.5K_{cu}$ | $\infty$ | 0 |
| PI | $0.45K_{cu}$ | $0.8T_u$ | 0 |
| PID | $0.6K_{cu}$ | $0.5T_u$ | $0.125T_u$ |

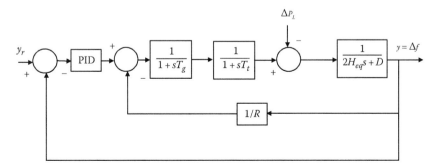

**FIGURE 4.9**
Isolated power area with a PID controller.

The plant transfer function is given by

$$\frac{Y}{U} = \frac{1}{\left(2H_{eq}s + D\right)\left(1 + sT_g\right)\left(1 + sT_t\right) + \left(1/R\right)} \tag{4.20}$$

The roots of the characteristic equation of this system are (−6.4146, −0.9802 ± $j$1.8649). Therefore, the step response will not be of S-shape due to the complex conjugate poles, and hence the first method cannot be applied. However, if we examine Figure 4.9, we see that the PID controller is in parallel with the droop characteristic $1/R$ and hence they can be combined together in an equivalent controller

$$G_e(s) = K_c\left(1 + \frac{1}{T_i s} + T_d s\right) + \frac{1}{R} \tag{4.21}$$

In this case, the plant transfer function without droop characteristic is given by

$$\frac{Y}{U} = \frac{1}{\left(2H_{eq}s + D\right)\left(1 + sT_g\right)\left(1 + sT_t\right)}$$

and its unit step response can be approximated by a FOPTD model (4.19). The values of $K$, $L$, and $T$ are found from the unit step response as $K = 2.0$, $L = 0.4$, and $T = 25$. The unit step response of the plant without droop characteristic and the approximated FOPTD (4.19) are shown in Figure 4.10.

From Table 4.1, the equivalent PID controller parameters are calculated as $K_c + (1/R) = 1.2(T/L) = 75$, $T_i = 2L = 0.8$, and $T_d = 0.5L = 0.2$. The controller gain is $K_c = 75 - (1/R) = 55$. The closed-loop system response due to a load increase of 20% is given in Figure 4.11.

   2. *Design using the second method of Ziegler–Nichols*

The closed-loop transfer function with only a proportional controller is given by

$$\frac{Y}{Y_r} = \frac{K_{cu}}{\left(2H_{eq}s + D\right)\left(1 + sT_g\right)\left(1 + sT_t\right) + \frac{1}{R} + K_{cu}} \tag{4.22}$$

where $K_{cu}$ is the gain that produces sustained oscillation. This value can be determined using Routh's stability criterion [7]. The closed-loop characteristic equation is given by

$$s^3 + 8.375s^2 + 17.014s + 28.47 + 1.389K_{cu} = 0$$

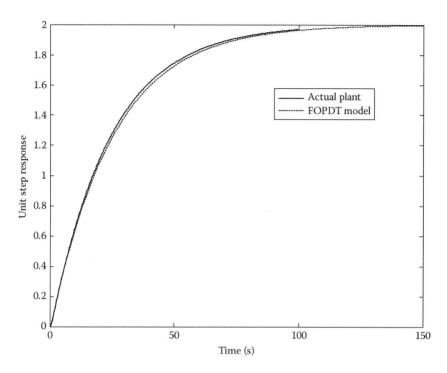

**FIGURE 4.10**
Unit step response of the single-area LFC and its FOPTD model.

The following Routh table is constructed

$$
\begin{array}{ccc}
s^3 & 1 & 17.014 \\
s^2 & 8.375 & 28.47 + 1.389K_{cu} \\
s^1 & 17.014 - \dfrac{28.47 + 1.389K_{cu}}{8.375} & \\
s^0 & 28.47 + 1.389K_{cu} &
\end{array}
$$

The sustained oscillation will occur when all coefficients in the $s^1$ row are zero. This gives $K_{cu} = 82.1$. The frequency of sustained oscillation is determined by solving the following auxiliary equation: $A(s) = 8.375s^2 + (28.47 + 1.389K_{cu}) = 0$ which gives $\omega_u = 4.125$ rad/s and hence $T_u = 1.52$ s. From Table 4.2, the PID controller parameters are calculated as $K_c = 0.6K_{cu} = 49.26$, $T_i = 0.5T_u = 0.76$, and $T_d = 0.125T_u = 0.19$. The closed-loop system response due to load change of 20% is depicted in Figure 4.12.

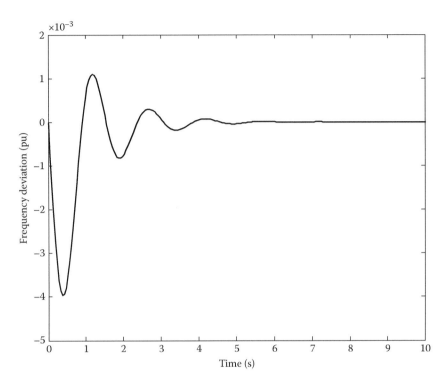

**FIGURE 4.11**
Frequency deviation of the closed-loop single area LFC with a PID controller designed using the first method of Ziegler–Nichols.

### 4.4.2 Analytical PID Design Technique

The internal model principle is a general analytical method for the design of control systems that can be applied to the PID controller design. The internal model control (IMC) was introduced by Garcia and Morari [8]. The IMC philosophy relies on the internal model principle, which states that accurate control can be achieved only if the controller captures some representation of the process to be controlled. To illustrate this concept, consider a simple open-loop control system, as shown in Figure 4.13. Suppose that $\hat{G}(s)$ is a model of the plant transfer function $G(s)$. If the controller $C(s)$ is selected as $1/\left(\hat{G}(s)\right)$, then

$$Y(s) = \frac{G(s)}{\hat{G}(s)} Y_d(s)$$

and perfect tracking occurs if $G(s) = \hat{G}(s)$.

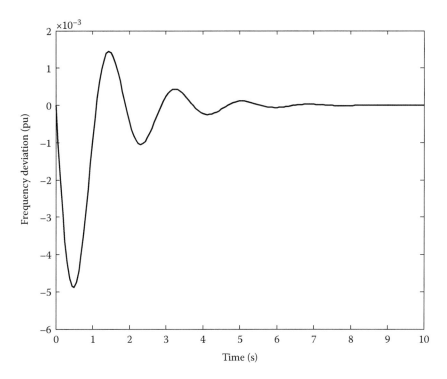

**FIGURE 4.12**
Frequency deviation of the closed-loop single area LFC with a PID controller designed using the second method of Ziegler–Nichols.

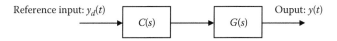

**FIGURE 4.13**
Open-loop control system.

This perfect tracking controller cannot be implemented in practice due to the following reasons:

1. The model $\hat{G}(s)$ does not accurately represent the plant $G(s)$, that is, there is always mismatch between the plant and its approximate model.
2. The model $\hat{G}(s)$ may not be invertible.
3. The plant is often subjected to unknown disturbances.

Therefore, the open-loop control structure will not be able to achieve tracking to a reference input. To overcome these difficulties, the IMC strategy

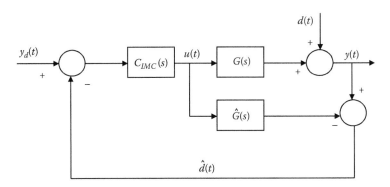

**FIGURE 4.14**
General structure of IMC.

having the general structure shown in Figure 4.14 is used. From this figure, it is clear that the plant and the model outputs are compared, resulting in the signal $\hat{d}(t)$. Note that $\hat{D}(s)$ can be written as

$$\hat{D}(s) = \left(G(s) - \hat{G}(s)\right)U(s) + D(s) \tag{4.23}$$

where $D(s)$ is the Laplace transform of $d(t)$ that represents an unknown disturbance input. If $D(s) = 0$, then $\hat{D}(s)$ can be viewed as the measure of the difference in behavior between the plant and its model. Therefore, $\hat{D}(s)$ can be used to improve the control by subtracting it from the reference input $Y_d(s)$. The resulting control signal $U(s)$ is written as

$$U(s) = C_{IMC}(s)\left(Y_d(s) - \hat{D}(s)\right) \tag{4.24}$$

Substituting (4.23) into (4.24), we get

$$U(s) = \frac{C_{IMC}(s)}{1 + C_{IMC}(s)\left(G(s) - \hat{G}(s)\right)}\left(Y_d(s) - D(s)\right) \tag{4.25}$$

Also, the closed-loop system output $Y(s)$ is determined as $Y(s) = G(s)U(s) + D(s)$, which, upon using (4.25), gives

$$Y(s) = T(s)Y_d(s) + S(s)D(s) \tag{4.26}$$

where

$$S(s) = \frac{1 - \hat{G}(s)C_{IMC}(s)}{1 + C_{IMC}(s)\left(G(s) - \hat{G}(s)\right)} \tag{4.27}$$

$$T(s) = \frac{G(s)C_{IMC}(s)}{1 + C_{IMC}(s)\left(G(s) - \hat{G}(s)\right)} \tag{4.28}$$

The functions $S(s)$ and $T(s)$ are termed as the sensitivity function and the complementary sensitivity function, respectively. In the absence of plant–model mismatch, that is, $G(s) = \hat{G}(s)$, and if

$$C_{IMC}(s) = \frac{1}{\hat{G}(s)} \tag{4.29}$$

then $S(s) = 0$ and $T(s) = 1$. This means that perfect reference input tracking and disturbance rejection are achieved. The controller (4.29) is known as the perfect tracking controller. In this case, (4.26) becomes

$$Y(s) = Y_d(s) \tag{4.30}$$

for all $t > 0$ and all disturbances $d(t)$. Detailed advantages and properties of the IMC structure shown in Figure 4.14 are reported in [9].

### 4.4.2.1 Practical Consideration of IMC Design

In practice, the perfect tracking controller given by (4.29) cannot be implemented due to several reasons [9]:

1. If the model $\hat{G}(s)$ has a positive zero (i.e., nonminimum phase), then the controller $C_{IMC}(s) = 1/\left(\hat{G}(s)\right)$ has a positive pole and hence the closed-loop system will be unstable.
2. If the model has a time delay term $e^{-\theta s}$, where $\theta$ the delay time in s, then the controller is predictive.
3. Moreover, if the model is strictly proper (i.e., the degree of the numerator is strictly less than the degree of the denominator), then the controller will be improper implying that for high frequency noise, the amplitude of the control signal will be very large.

The remedy of these realization problems is given in the following design procedure for a practical IMC controller.

*Procedure to design a practical IMC*

Step 1:

Factorize the model as

$$\hat{G}(s) = \hat{G}_+(s)\hat{G}_-(s) \tag{4.31}$$

where the factor $\hat{G}_+(s)$ contains all time delays and positive zeros and the factor $\hat{G}_-(s)$ has no delays and all of its zeros are negative.

Step 2:

Define the IMC controller as

$$C_{IMC}(s) = \frac{1}{\hat{G}_-(s)}F(s) \tag{4.32}$$

where $F(s)$ is a low-pass filter chosen such that $C_{IMC}(s)$ is proper. The simplest filter takes the form

$$F(s) = \frac{1}{(1 + T_f s)^r} \tag{4.33}$$

where

$r$ is selected to guarantee the properness of $C_{IMC}(s)$

$T_f$ is an adjustable parameter used to determine the speed of response

Using the IMC controller given by (4.32), the sensitivity and the complementary sensitivity functions defined by (4.27) and (4.28) become

$$S(s) = \frac{1 - \hat{G}(s)\hat{G}_-^{-1}(s)F(s)}{1 + \hat{G}_-^{-1}(s)F(s)\left(G(s) - \hat{G}(s)\right)} \tag{4.34}$$

$$T(s) = \frac{G(s)\hat{G}_-^{-1}(s)F(s)}{1 + \hat{G}_-^{-1}(s)F(s)\left(G(s) - \hat{G}(s)\right)} \tag{4.35}$$

In this case, the closed-loop output is given by

$$Y(s) = \frac{G(s)\hat{G}_-^{-1}(s)F(s)}{1 + \hat{G}_-^{-1}(s)F(s)\left(G(s) - \hat{G}(s)\right)}Y_d(s) + \frac{1 - \hat{G}(s)\hat{G}_-^{-1}(s)F(s)}{1 + \hat{G}_-^{-1}(s)F(s)\left(G(s) - \hat{G}(s)\right)}D(s) \tag{4.36}$$

When there is no plant–model mismatch, then (4.36) becomes

$$Y(s) = \hat{G}(s)\hat{G}_-^{-1}(s)F(s)Y_d(s) + \left(1 - \hat{G}(s)\hat{G}_-^{-1}(s)F(s)\right)D(s) \tag{4.37}$$

Using (4.31) in (4.37), we get

$$Y(s) = \hat{G}_+(s)F(s)Y_d(s) + \left(1 - \hat{G}_+(s)F(s)\right)D(s) \tag{4.38}$$

**Example 4.2**

Design an IMC for the single-area LFC system with a hydro turbine shown in Figure 4.15. The system data are $T_W = 2$, $T_g = 0.5$, $R = 0.05$, $H = 5$, and $D = 1$.

**Solution:**

The plant transfer function is given by

$$\frac{Y(s)}{U(s)} = G(s) = \frac{(1 - T_w s)}{(2Hs + D)(1 + sT_g)(1 + 0.5sT_w) + (1/R)} \tag{4.39}$$

The model transfer function $\hat{G}(s)$ can be taken as the plant transfer function, that is, $\hat{G}(s) = G(s)$. Note that $G(s)$ is nonminimum phase, then $\hat{G}_+(s)$ and $\hat{G}_-(s)$ are given by

$$\hat{G}_+(s) = (1 - T_w s) \tag{4.40}$$

$$\hat{G}_-(s) = \frac{1}{(2Hs + D)(1 + sT_g)(1 + 0.5sT_w) + (1/R)} \tag{4.41}$$

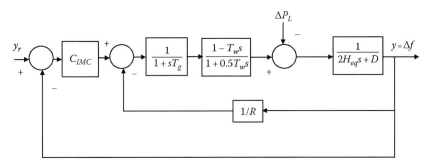

**FIGURE 4.15**
Single-area LFC with a hydro turbine unit.

Hence, the IMC controller (4.32) is determined as

$$C_{IMC}(s) = \frac{F(s)}{\hat{G}_-(s)} = \frac{(2Hs+D)(1+sT_g)(1+0.5sT_w)+(1/R)}{(1+T_f s)^3} \quad (4.42)$$

where the filter $F(s)$ is chosen of third order to ensure that $C_{IMC}(s)$ is proper. When the plant–model mismatch is zero and the reference input $Y_d(s)=0$, the closed-loop output and the control signal in the presence of the designed IMC controller can be determined from (4.38) and (4.25) as

$$Y(s) = \left(1 - G_+(s)F(s)\right)D(s) \quad (4.43)$$

$$U(s) = -D(s)G_-^{-1}(s)F(s) \quad (4.44)$$

Using the system data, the impact of this controller on the closed-loop frequency deviation due to 20% load change for different filter time constant ($T_f$=0.5, 1.0 and 2.0 s) is shown in Figure 4.16 and the control signal is depicted in Figure 4.17. It is clear that as $T_f$ increases, the output response becomes slower.

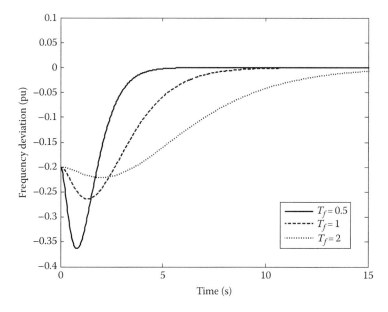

**FIGURE 4.16**
Closed-loop frequency deviation of a single-area LFC under the effect of IMC.

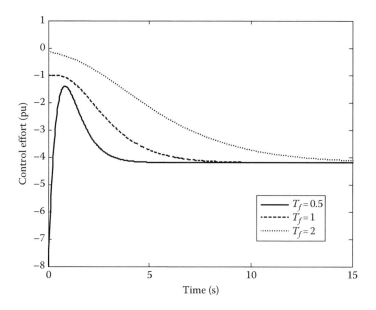

**FIGURE 4.17**
Control effort under the effect of IMC.

### 4.4.2.2 Relation between IMC and PID Controllers Design

The relation between the classical controller $C(s)$ and the IMC controller $C_{IMC}(s)$ can be found by equating the closed-loop transfer functions of the two disturbance-free systems shown in Figures 4.14 and 4.18. These transfer functions are given by

$$\frac{G(s)C_{IMC}(s)}{1+C_{IMC}(s)\left(G(s)-\hat{G}(s)\right)}=\frac{G(s)C(s)}{1+G(s)C(s)} \tag{4.45}$$

Solving (4.45) to get the following relation between $C(s)$ and $C_{IMC}(s)$

$$C(s)=\frac{C_{IMC}(s)}{1-C_{IMC}(s)\hat{G}(s)} \tag{4.46}$$

Substituting (4.31) and (4.32) into (4.46) yields

$$C(s)=\frac{\left(1/\left(\hat{G}_-(s)\right)\right)F(s)}{1-F(s)\hat{G}_+(s)}=\frac{\hat{G}_-^{-1}(s)}{F^{-1}-\hat{G}_+(s)} \tag{4.47}$$

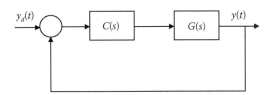

**FIGURE 4.18**
Closed-loop control system with a classical controller.

Note that if the plant is minimum phase, that is, $\hat{G}_+(s) = 1$ and the filter $F(s)$ is of first order, then (4.47) reduces to

$$C(s) = \frac{\hat{G}_-^{-1}(s)}{T_f s} \qquad (4.48)$$

### Example 4.3

Consider a single-area LFC system with non-reheat turbine unit. A block diagram of such a system is shown in Figure 4.19, where $T_t = 0.2$, $T_g = 0.3$, $H = 6$, $D = 0.5$, and $R = 0.05$.

1. Design the IMC controller $C_{IMC}(s)$.
2. Convert $C_{IMC}(s)$ to an equivalent PID controller.

**Solution:**

1. In this example, the plant transfer function takes the form

$$G(s) = \frac{1}{(2Hs + D)(1 + sT_t)(1 + sT_g) + (1/R)} \qquad (4.49)$$

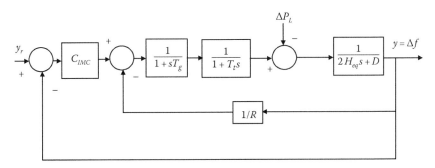

**FIGURE 4.19**
Single-area LFC with a non-reheat turbine unit.

This transfer function can be approximated by a second-order model using the dominate poles. For the numerical data, the characteristic equation of $G(s)$ is calculated as $(s^3 + 8.37s^2 + 17.01s + 28.47)$ with the roots at $(-6.415, -0.98, \pm j1.865)$. Since the real part of the complex conjugate roots is less than one-fifth of the real root, the real root can be neglected [10]. The second-order approximation model of the plant takes the form

$$\hat{G}(s) = \frac{K}{\left(s^2 + 1.96s + 4.44\right)} \tag{4.50}$$

where $K$ is chosen as $K = 4.44/(D + (1/R)) = 0.216$ such that the DC gains of $G(s)$ and $\hat{G}(s)$ are equal. The step responses of both $G(s)$ and $\hat{G}(s)$ are given in Figure 4.20. These responses validate the dominant poles approximation. Since $\hat{G}(s)$ is the minimum phase, $\hat{G}_+(s) = 1$ and $\hat{G}_-(s) = \hat{G}(s)$. The IMC controller is designed as

$$C_{IMC}(s) = \hat{G}^{-1}F(s) = \frac{\left(s^2 + 1.96s + 4.44\right)}{K\left(1 + T_f s\right)^r} \tag{4.51}$$

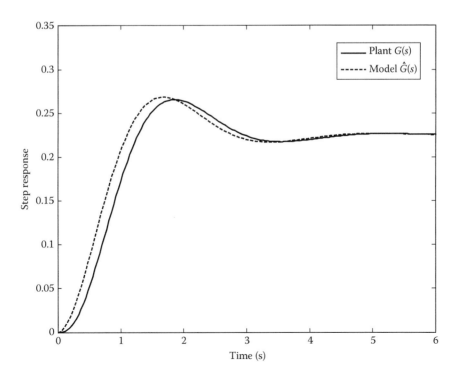

**FIGURE 4.20**
Comparison between step responses of the plant $G(s)$ and the model $\hat{G}(s)$.

2. Using (4.47), the controller (4.51) can be converted to an equivalent classical controller $C(s)$

$$C(s) = \frac{\left(s^2 + 1.96s + 4.44\right)}{K\left(\left(1 + T_f s\right)^r - 1\right)} \qquad (4.52)$$

where $K = 0.216$. If $T_f = 1$ and $r = 1$, then (4.52) becomes the following ideal PID controller:

$$C(s) = \frac{\left(s^2 + 1.96s + 4.44\right)}{0.216s} \qquad (4.53)$$

Comparing (4.53) and (4.2), we obtain the PID parameters as $K_c = 9.07$, $T_i = 0.44$, and $T_d = 0.51$. For $T_f = 1$ and $r = 2$, (4.52) becomes

$$C(s) = \frac{\left(s^2 + 1.96s + 4.44\right)}{0.432s(1 + 0.5s)} \qquad (4.54)$$

The controller (4.54) is in the form of an ideal PID controller cascaded with a first-order filter given by (4.16). Comparing (4.54) and (4.16), we obtain the following controller parameters: $K_c = 4.53$, $T_i = 0.44$, and $T_d = 0.51$. The first-order filter time constant is 0.5 s.

## 4.5 Model Reduction and Tuning of PID Controllers

As has been explained in Example 4.3, the key point in designing a PID controller using the IMC principle is to find a reduced-order model of the plant in order to deduce a direct relation between the IMC and the PID controllers. Indeed, there are various techniques for model reduction and PID tuning. In this section, the two-step procedure technique to find a FOPTD and a second-order plus time delay (SOPTD) models of a given open-loop transfer function [11] is presented.

### 4.5.1 The Two-Step Technique

The first step of this technique is to determine a model of the plant $\hat{G}(s)$ using the half rule technique [10]. The second step is to use the model $\hat{G}(s)$ to find the IMC-based PID controller parameters as demonstrated in the previous example.

### 4.5.1.1 The Half Rule Approximation Technique

Consider a nonminimum phase plant having the general transfer function form

$$G(s) = e^{-\theta_o s} \frac{\prod_j (1 - \bar{T}_{jo}s) \prod_k (1 + T_{ko}s)}{\prod_i (1 + \tau_{io}s)} \tag{4.55}$$

where $\tau_{io}$, $\bar{T}_{jo}$, $T_{ko}$, and $\theta_o$ are positive parameters and the time constants $\tau_{io}$ are arranged in descending order. The half rule gives a simple formula for approximating (4.55) to a FOPTD or a SOPTD transfer function of the form

$$\hat{G}(s) = \frac{1}{1 + \tau_1 s} e^{-\theta s} \tag{4.56}$$

$$\hat{G}(s) = \frac{1}{(1 + \tau_1 s)(1 + \tau_2 s)} e^{-\theta s} \tag{4.57}$$

If all $T_{ko} = 0$, then the FOPTD approximation rules are given by

$$\left. \begin{aligned} \tau_1 &= \tau_{1o} + 0.5\tau_{2o} \\ \theta &= \theta_o + 0.5\tau_{2o} + \sum_{i \geq 3} \tau_{io} + \sum_j \bar{T}_{jo} \end{aligned} \right\} \tag{4.58}$$

and the SOPTD approximation rules are given by

$$\left. \begin{aligned} \tau_1 &= \tau_{1o} \\ \tau_2 &= \tau_{2o} + 0.5\tau_{3o} \\ \theta &= \theta_o + 0.5\tau_{3o} + \sum_{i \geq 4} \tau_{io} + \sum_j \bar{T}_{jo} \end{aligned} \right\} \tag{4.59}$$

When some of the positive numerator time constants $T_{ko}$ are not zero, it is proposed [10] to cancel them against the neighboring dominator time constants one at a time starting from the largest $T_{ko}$ according to

$$\frac{1 + T_o s}{1 + \tau_o s} = \frac{1}{1 + (\tau_o - T_o)s} \tag{4.60}$$

**Example 4.4**

Find a FOPTD approximate model for the plant

$$G(s) = \frac{(1+4s)(1+1.2s)(1-s)}{(1+8s)(1+5s)(1+1.5s)(1+0.5s)(1+0.2s)}$$

**Solution:**

First, we approximate the positive numerator time constants $T_{1o}=4$ and $T_{2o}=1.2$ one at a time starting from $T_{1o}=4$. This can be done by selecting the closest larger denominator time constant, which is $\tau_o=5$. Therefore, $(1+4s)/(1+5s)$ is approximated according to (4.60) as $1/(1+s)$. The same procedure is applied to $T_{2o}=1.2$ and $\tau_o=1.5$, which yields $(1+1.2s)/(1+1.5s)\approx 1/(1+0.3s)$. The given transfer function can now be approximated as

$$\bar{G}(s) \simeq \frac{(1-s)}{(1+8s)(1+s)(1+0.5s)(1+0.3s)(1+0.2s)}$$

Note, according to (4.55), we have $\tau_{1o}=8$, $\tau_{2o}=1$, $\tau_{3o}=0.5$, $\tau_{4o}=0.3$, $\tau_{5o}=0.2$, $\bar{T}_{1o}=1$, and $\theta_o=0$. The last step is to calculate the parameters of the FOPTD model from (4.58) as $\tau_1=\tau_{1o}+0.5\tau_{2o}=8.5$ and $\theta=0.5\tau_{2o}+\tau_{3o}+\tau_{4o}+\tau_{5o}+\bar{T}_{1o}=2.5$. The approximate FOPTD model is then given by $\hat{G}(s)=1/(1+8.5s)e^{-2.5s}$. Simulation results of both the actual plant $G(s)$ and its FOPTD model $\hat{G}(s)$ are shown in Figure 4.21. These results validate the half rule approximation technique.

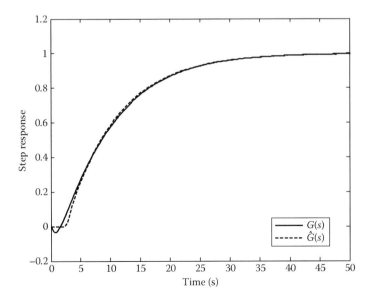

**FIGURE 4.21**
Comparison between step responses of $G(s)$ and $\hat{G}(s)$ of Example 4.4.

**Example 4.5**

The purpose of this example is to find a FOPTD and a SOPTD model using the half rule method for the single-area LFC system shown in Figure 4.22 with

1. A non-reheat turbine
2. A reheat turbine
3. A hydro turbine

The system parameters are given in Table 4.3.

The plant transfer functions of the single-area LFC system with different types of turbine are given by

$$G_{non}(s) = \frac{(1/D)}{(1+(2H/D)s)(1+sT_t)(1+sT_g)}$$

$$(\text{non-reheat turbine}) \tag{4.61}$$

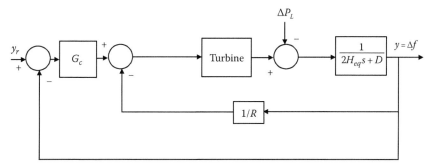

**FIGURE 4.22**
Single-area LFC with a different turbine unit.

**TABLE 4.3**

Parameters of a Single-Area LFC with Different Types of Turbines

| Parameters | Non-Reheat | Reheat | Hydro |
|---|---|---|---|
| $H$ | 6.0 | 5.0 | 5.0 |
| $D$ | 0.5 | 0.25 | 1.0 |
| $T_g$ | 0.3 | 0.25 | 0.5 |
| $T_t$ | 0.2 | 0.3 | |
| $T_r$ | | 7.0 | |
| $K_t$ | | 0.3 | |
| $T_W$ | | | 2.0 |

$$G_r(s) = \frac{(1/D)(1+sK_tT_r)}{(1+(2H/D)s)(1+sT_r)(1+sT_t)(1+sT_g)}$$

$$\text{(reheat turbine)} \tag{4.62}$$

$$G_h(s) = \frac{(1/D)(1-sT_w)}{(1+(2H/D)s)(1+sT_g)(1+0.5sT_w)}$$

$$\text{(hydro turbine)} \tag{4.63}$$

Substituting the data in Table 4.3, these transfer functions become

$$G_{non}(s) = \frac{2}{(1+24s)(1+0.3s)(1+0.2s)} \tag{4.64}$$

$$G_r(s) = \frac{4(1+2.1s)}{(1+40s)(1+7s)(1+0.3s)(1+0.25s)} \tag{4.65}$$

$$G_h(s) = \frac{(1-2s)}{(1+10s)(1+0.5s)(1+s)} \tag{4.66}$$

It is worth mentioning that in these transfer functions, the governor droop characteristic $1/R$ is not considered. However, in the controller design stage, it can be considered by adding it to the controller $G_c(s)$ [12]. The procedure of the half rule method is followed to determine the FOPTD and SOPTD models for each of the transfer functions (4.64) through (4.66). In applying the half rule procedure for the transfer function (4.65), the term $(1+2.1s)/(1+7s)$ is approximated by $1/(1+4.9s)$. Therefore, the model (4.65) with an approximation of the positive numerator time constant is given by

$$\overline{G}_r(s) = \frac{4}{(1+40s)(1+4.9s)(1+0.3s)(1+0.25s)} \tag{4.67}$$

The parameters of the FOPTD model are then calculated as

$$\tau_1 = \tau_{1o} + 0.5\tau_{2o} = 40 + 0.5 \times 4.9 = 42.45$$

$$\theta = 0.5\tau_{2o} + \tau_{3o} + \tau_{4o} = 0.5 \times 4.9 + 0.3 + 0.25 = 3$$

The FOPTD model is given by

$$\hat{G}_r(s) = \frac{4}{1+42.45s}e^{-3s}$$

The parameters of the SOPTD model are calculated using (4.59) as

$$\tau_1 = \tau_{1o} = 40$$
$$\tau_2 = \tau_{2o} + 0.5\tau_{3o} = 4.9 + 0.5 \times 0.3 = 5.05$$

$$\theta = 0.5\tau_{3o} + \tau_{4o} = 0.5 \times 0.3 + 0.3 + 0.25 = 0.4$$

The SOPTD model is given by

$$\hat{G}_r(s) = \frac{4}{(1+40s)(1+5.05s)} e^{-0.4s}$$

The FOPTD and SOPTD models of all the transfer functions given in Equations 4.64 through 4.66 are shown in Table 4.4.

Validation of the approximate models and the actual plants using the open-loop step responses are shown in Figures 4.23 through 4.25.

### Example 4.6

For the LFC system with a hydro turbine considered in Example 4.5, design PI and PID controllers using the FOPTD and SOPTD models.

*PI controller design*

The FOPTD model obtained in Example 4.5 is used to design the PI controller. From the model

$$\hat{G}_h(s) = \frac{1}{1+10.5s} e^{-3s}$$

**TABLE 4.4**

FOPTD and SOPDT Models of Single-Area LFC Systems with Different Types of Turbines

| Plant | FOPTD | SOPTD |
|---|---|---|
| $G_{non}(s) = \dfrac{2}{(1+24s)(1+0.3s)(1+0.2s)}$ | $\hat{G}_{non} = \dfrac{2}{1+24.15s} e^{-0.35s}$ | $\hat{G}_{non} = \dfrac{2}{(1+24s)(1+0.4s)} e^{-0.1s}$ |
| $G_r(s) = \dfrac{4(1+2.1s)}{(1+40s)(1+7s)(1+0.3s)(1+0.25s)}$ | $\hat{G}_r(s) = \dfrac{4}{1+42.45s} e^{-3s}$ | $\hat{G}_r(s) = \dfrac{4}{(1+40s)(1+5.05s)} e^{-0.4s}$ |
| $G_h(s) = \dfrac{(1-2s)}{(1+10s)(1+0.5s)(1+s)}$ | $\hat{G}_h(s) = \dfrac{1}{1+10.5s} e^{-3s}$ | $\hat{G}_h(s) = \dfrac{1}{(1+10s)(1+1.25s)} e^{-2.25s}$ |

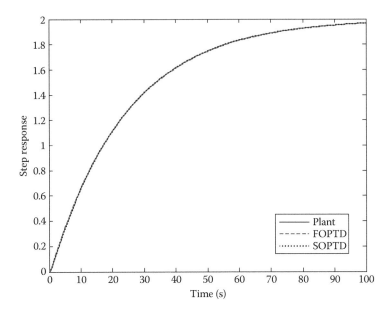

**FIGURE 4.23**
Step response of non-reheat-type system and its approximate models.

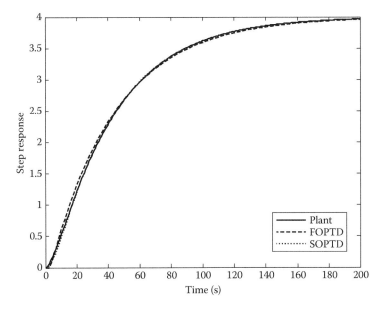

**FIGURE 4.24**
Step response of reheat-type system and its approximate models.

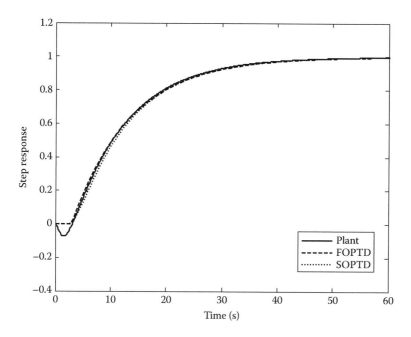

**FIGURE 4.25**
Step response of hydro-type system and its approximate models.

The factorization terms $\hat{G}_{h-}(s)$ and $\hat{G}_{h+}(s)$ are given by

$$\hat{G}_{h-}(s) = \frac{1}{1+10.5s}$$

$$\hat{G}_{h+}(s) = e^{-3s}$$

Use (4.47) to find the following controller:

$$C(s) = \frac{\hat{G}_{h-}^{-1}(s)}{F^{-1} - \hat{G}_{h+}(s)} = \frac{1+10.5s}{\left(1+T_f s\right) - e^{-3s}} \tag{4.68}$$

To show that this controller is in fact a PI controller, the term $e^{-3s}$ is approximated using Taylor series expansion as $e^{-3s} \simeq (1-3s)$ and the controller (4.68) takes the form

$$C(s) = \frac{1+\tau_1 s}{s\left(T_f + 3\right)} \tag{4.69}$$

When $T_{d1} = 0$, the controller (4.3) becomes a PI controller of the form

$$G_s(s) = K_{c1}\left(1 + \frac{1}{T_{i1}s}\right) \tag{4.70}$$

Comparing (4.69) with (4.70) shows that the designed controller $C(s)$ is equivalent to a PI controller with the following parameters:

$$\left.\begin{array}{l} K_{c1} = \dfrac{\tau_1}{T_f + \theta} = \dfrac{10.5}{T_f + 3} \\[2mm] T_{i1} = \tau_1 = 10.5 \end{array}\right\} \tag{4.71}$$

*PID controller design*
  The SOPTD model

$$\hat{G}_h(s) = \frac{1}{(1+10s)(1+1.25s)}e^{-2.25s}$$

is used to design a PID controller. In this case, the factorization of $\hat{G}_{h-}(s)$ and $\hat{G}_{h+}(s)$ will be

$$\hat{G}_{h-}(s) = \frac{1}{(1+10s)(1+1.25s)}$$

$$\hat{G}_{h+}(s) = e^{-2.25s}$$

Using (4.47), the controller $C(s)$ takes the form

$$C(s) = \frac{(1+10s)(1+1.25s)}{\left((1+T_f s)^r - e^{-2.25s}\right)} \tag{4.72}$$

Indeed, this controller is a PID controller. To see this, replace the term $e^{-2.25s}$ by its first-order Taylor series expansion $(1-2.25s)$ and then writing (4.72) when $r=1$ as

$$C(s) = \frac{(1+10s)(1+1.25s)}{s(T_f + 2.25)} \tag{4.73}$$

and when $r=2$ as

$$C(s) = \frac{(1+10s)(1+1.25s)}{s\left(2T_f + 2.25\right)\left(1+\left(T_f^2/\left(2T_f + 2.25\right)\right)s\right)}$$ (4.74)

In fact, (4.73) and (4.74) are PID controllers. The PID parameters of (4.73) are obtained, by comparing (4.73) with (4.3), as

$$\left.\begin{aligned} K_{c1} &= \frac{10}{\left(T_f + 2.25\right)} \\ T_{i1} &= 10 \\ T_{d1} &= 1.25 \end{aligned}\right\}$$ (4.75)

Similarly, the PID parameters of (4.74) are obtained, by comparing (4.74) with (4.15), as

$$\left.\begin{aligned} K_{c1} &= \frac{10}{\left(2T_f + 2.25\right)} \\ T_{i1} &= 10 \\ T_{d1} &= 1.25 \\ N &= \frac{1.25}{T_f^2}\left(2T_f + 2.25\right) \end{aligned}\right\}$$ (4.76)

Therefore, the PID controllers of (4.73) and (4.74) take the following forms, respectively:

$$G_s(s) = \frac{10}{T_f + 2.25}\left(1 + \frac{1}{10s}\right)(1+1.25s)$$ (4.77)

$$G_{cm}(s) = K_{c1}\left(1 + \frac{1}{10s}\right)\left(\frac{1+1.25s}{1+(1.25/N)s}\right)$$ (4.78)

where $K_{c1}$ and $N$ are as defined in (4.76). The frequency deviations of the system under load disturbance of 20% are simulated for the PI controller (4.70) and (4.71) and the PID controllers (4.77) and (4.78) when $T_f=3.75$. The simulation results are shown in Figure 4.26 and the control effort for all controllers is presented in Figure 4.27.

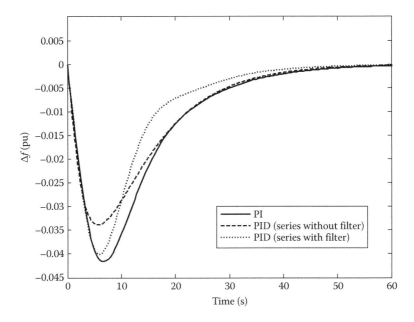

**FIGURE 4.26**
Frequency deviations for PI and PID controllers.

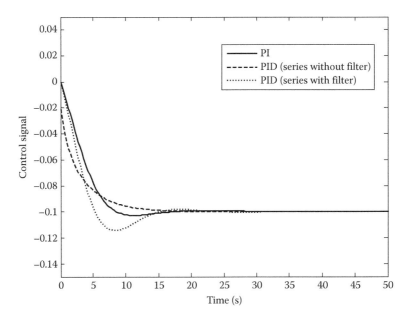

**FIGURE 4.27**
Control signals for PI and PID controllers.

# References

1. J. G. Ziegler and N. B. Nichols, Optimum settings for automatic controllers, *Trans. ASME*, 64, 759–768, 1942.
2. A. O'Dwyer, *Handbook of PI and PID Controller Tuning Rules*, 2nd edn., Imperial College Press, London, U.K, 2006.
3. K. J. Åström and T. Hägglund, *PID Controllers: Theory, Design and Tuning*, Instrument Society of America, Research Triangle Park, NC, 1995.
4. A. Visioli, *Practical PID Control*, Springer-Verlag, London, U.K., 2006.
5. K. J. Åström and T. Hägglund, Revisiting the Ziegler–Nichols step response method for PID control, *J. Process Control*, 14, 635–650, 2004.
6. K. H. Ang, G. Chong, and Y. Li, PID control systems analysis, design, and technology, *IEEE Trans. Control Syst. Technol.*, 13, 559–576, 2005.
7. K. Ogata, *Modern Control Engineering*, 4th edn., Prentice Hall, Upper Saddle River, NJ, 2002.
8. C. E. Garcia and M. Morari, Internal model control. A unifying review and some new results, *Ind. Eng. Chem. Process Des. Dev.*, 21(2), 308–323, 1982.
9. D. E. Rivera, M. Morari, and S. Skogestad, Internal model control. 4. PID controller design, *Ind. Eng. Chem. Process Des. Dev.*, 25, 252–265, 1986.
10. N. S. Nise, *Control Systems Engineering*, 5th edn., John Wiley & Sons, New York, 2008.
11. S. Skogestad, Simple analytic rules for model reduction and PID controller tuning, *J. Process Control*, 13, 291–309, 2003.
12. W. Tan, Unified tuning of PID load frequency controller for power systems via IMC, *IEEE Trans. Power Syst.*, 25(1), 341–350, 2010.

# 5

# *Decentralized LFC Design for a Multi-Area Power System*

## 5.1 Introduction

The LFC objectives in a multi-area power system are to keep, simultaneously, the frequency deviation of each area at the nominal values and the tie-line power changes at the scheduled levels. To achieve these objectives, the area control error (ACE) introduced in Chapter 1 is used as a feedback signal. A decentralized LFC design for a multi-area power system is presented in this chapter. This methodology is based on the design of individual controller for an isolated area system. Then, the stability of the overall interconnected system is checked to ensure that the independently designed local controllers achieve the desired performance. Closed-loop stability is investigated using structured singular value (SSV) of the interaction between the entries of the open-loop transfer function matrix.

## 5.2 Decentralized PID Controllers

Decentralized (diagonal) multivariable controllers are often preferred over fully cross-coupled multivariable controllers since the former have simpler structure and include less number of tuning parameters. Consider the multivariable control system shown in Figure 5.1 where $G(s)$ is an $m \times m$ plant transfer function matrix, $C(s)$ is an $m \times m$ controller transfer function matrix, and $R(s), E(s), U(s)$, and $Y(s)$ are $m \times 1$ reference, error, control, and output vectors, respectively.

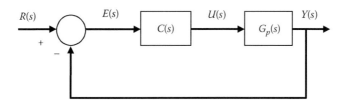

**FIGURE 5.1**
Multivariable control system.

The multivariable controller equation takes the form

$$U(s) = C(s)E(s) \tag{5.1}$$

In a decentralized control scheme, (5.1) will have the following diagonal structure:

$$
\begin{bmatrix} U_1(s) \\ U_2(s) \\ \vdots \\ U_m(s) \end{bmatrix} = \begin{bmatrix} C_1(s) & 0 & \cdots & 0 \\ 0 & C_2(s) & 0 & 0 \\ \vdots & \vdots & \ddots & \vdots \\ 0 & 0 & \cdots & C_m(s) \end{bmatrix} \begin{bmatrix} E_1(s) \\ E_2(s) \\ \vdots \\ E_m(s) \end{bmatrix} \tag{5.2}
$$

In block diagram form, decentralized multivariable control has a simple loop structure, as shown in Figure 5.2 ($m=2$).

A decentralized control of the form (5.2) will be designed for a multi-area LFC system. The design is based on model reduction and controller tuning presented in Chapter 4. In this scheme, the IMC-PI or PID local controller is designed using FOPTD or SOPTD models respectively. Stability of the closed-loop system with these local controllers is then studied. To this effect,

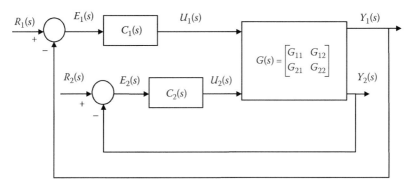

**FIGURE 5.2**
Diagonal decentralized control structure for $m = 2$.

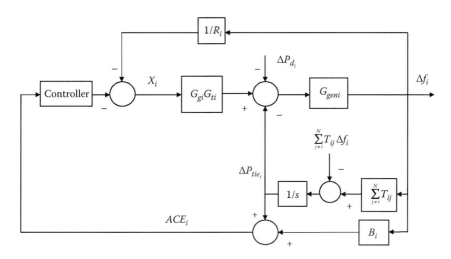

**FIGURE 5.3**
LFC scheme of the *i*th area in a multi-area power system.

consider an *N*-area LFC system. The block diagram of the *i*th area is shown in Figure 5.3. The area control error $ACE_i$ of each area is given by

$$ACE_i = \Delta P_{tie_i} + B_i \Delta f_i \quad i = 1,\ldots, N \tag{5.3}$$

In this figure, the transfer function of the generator is written in the form

$$G_{geni}(s) = \frac{K_{geni}}{1 + T_{geni}s}$$

where
$$K_{geni} = 1/D_i$$
$$T_{geni} = 2H_i/D_i$$

Each control area may have different types of turbines, for example, non-reheat, reheat, or hydro turbines.

Letting $\Delta P_{tie_i} = 0$, the local LFC signal $U_i$ can be expressed as

$$U_i = C_i B_i \Delta f_i = \overline{C}_i \Delta f_i \tag{5.4}$$

where $\overline{C}_i = C_i B_i$ is the local LFC controller, assuming that tie-line power is zero. Note also that the input signal to the governor $X_i$ in Figure 5.3 can be written as $X_i = -(\overline{C}_i + 1/R_i)\Delta f_i$. This suggests that the governor droop characteristic $(1/R_i)$ can be ignored in the stage of model reduction and design and then considered later by decreasing the proportional gain of the final controller by $(1/R_i)$ [1].

Without loss of generality, consider the *i*th area with a non-reheat turbine, then the plant transfer function can be written as (see Equation 4.61)

$$G_{noni}(s) = \frac{K_{geni}}{(1 + T_{geni}s)(1 + T_{ti}s)(1 + T_{gi}s)} \tag{5.5}$$

By using the model reduction technique presented in Chapter 4, the local plant (5.5) can be modeled in one of the following forms:

$$\hat{G}_{Fi}(s) = \frac{K_{geni}}{1 + \tau_1 s} e^{-\theta s} \tag{5.6}$$

$$\hat{G}_{Si}(s) = \frac{K_{geni}}{(1 + \tau_1 s)(1 + \tau_2 s)} e^{-\theta s} \tag{5.7}$$

The FOPTD model (5.6) is used to design a PI controller in the form

$$C_i = K_{c1}\left(1 + \frac{1}{T_{i1}s}\right) \tag{5.8}$$

where

$$K_{c1} = \frac{\tau_1}{(\theta + T_f)}$$

$$T_{i1} = \tau_1$$

The SOPTD model (5.7) is used to design a PID controller in the series form

$$C_i = K_{c1}\left(1 + \frac{1}{T_{i1}s}\right)(1 + T_{d1}s) \tag{5.9}$$

where

$$K_{c1} = \frac{\tau_1}{K(T_f + \theta)}$$

$$K = K_{geni}$$

$$T_{i1} = \tau_1$$

$$T_{d1} = \tau_2$$

Also, the SOPTD can be used to design a modified classical PID controller:

$$C_i = K_{c1}\left(1 + \frac{1}{T_{i1}s}\right)\frac{(1 + T_{d1}s)}{(1 + (T_{d1}/N)s)} \tag{5.10}$$

where

$$K_{c1} = \frac{\tau_1}{K(2T_f + \theta)}$$

$$K = K_{geni}$$
$$T_{i1} = \tau_1$$
$$T_{d1} = \tau_2$$
$$N = \frac{\tau_2}{T_f}\left(2 + \frac{\theta}{T_f}\right)$$

For details see Example 4.6. After designing the local controller for each area, the closed-loop stability of the overall system when tie-lines are in operation has to be examined.

The tie-line power deviation of each area can be written as

$$\Delta P_{tie_i} = \frac{2\pi}{s}\sum_{\substack{j=1 \\ j\neq i}}^{N} T_{ij}(\Delta f_i - \Delta f_j) \tag{5.11}$$

or in the following matrix form:

$$\begin{bmatrix} \Delta P_{tie1} \\ \Delta P_{tie2} \\ \vdots \\ \Delta P_{tieN} \end{bmatrix} = \frac{2\pi}{s} \begin{bmatrix} \sum_{j\neq 1}^{N} T_{1j} & -T_{12} & \cdots & -T_{1N} \\ -T_{21} & \sum_{j\neq 2}^{N} T_{2j} & \cdots & -T_{2N} \\ \vdots & \vdots & \vdots & \vdots \\ -T_{N1} & -T_{N2} & \cdots & \sum_{j\neq N}^{N} T_{Nj} \end{bmatrix} \begin{bmatrix} \Delta f_1 \\ \Delta f_2 \\ \vdots \\ \Delta f_N \end{bmatrix} \tag{5.12}$$

Defining the vectors $\Delta P_{tie} = [\Delta P_{tie1} \ \Delta P_{tie2} \cdots \Delta P_{tieN}]^T$ and $\Delta f = [\Delta f_1 \ \Delta f_2 \cdots \Delta f_N]^T$, then (5.12) can be written in the following compact form:

$$\Delta P_{tie} = \frac{2\pi}{s} T \Delta f \tag{5.13}$$

where

$$T = \begin{bmatrix} \sum_{j\neq 1}^{N} T_{1j} & -T_{12} & \cdots & -T_{1N} \\ -T_{21} & \sum_{j\neq 2}^{N} T_{2j} & \cdots & -T_{2N} \\ \vdots & \vdots & \vdots & \vdots \\ -T_{N1} & -T_{N2} & \cdots & \sum_{j\neq N}^{N} T_{Nj} \end{bmatrix}$$

is a square constant symmetric matrix (since $T_{ij} = T_{ji}$).

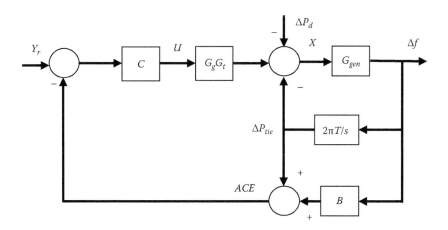

**FIGURE 5.4**
Multi-variable LFC scheme.

The block diagram shown in Figure 5.3 is redrawn as a multivariable block diagram form, as shown in Figure 5.4, where the transfer function matrices of the controllers, governors, turbines, and generators are diagonal matrices defined as $C = diag[C_i(s)]$, $G_g = diag[G_{gi}(s)]$, $G_t = diag[G_{ti}(s)]$, and $G_{gen} = diag[G_{geni}(s)]$, respectively. The vector $\Delta P_d$, the bias factor matrix $B$, and $ACE$ are defined as $\Delta P_d = [\Delta P_{d1} \ \Delta P_{d2} \cdots \Delta P_{dN}]^T$, $B = diag[B_i]$, and $ACE = [ACE_1 \ ACE_2 \cdots ACE_N]^T$.

The open-loop transfer function matrix $\overline{G}$ defined by

$$\Delta f = \overline{G}(-U) \tag{5.14}$$

can be found from the block diagram shown in Figure 5.4 by solving the following equations:

$$X = -\Delta P_d - \Delta P_{tie} - G_g G_t U \tag{5.15}$$

$$\Delta P_{tie} = \left(2\pi s^{-1} T\right) \Delta f \tag{5.16}$$

$$\Delta f = G_{gen} X \tag{5.17}$$

Substituting (5.16) into (5.15) and then plugging the results into (5.17), we get

$$\left(I + 2\pi s^{-1} G_{gen} T\right) \Delta f = G_{gen} G_g G_t \left(-U\right) + G_{gen} \left(-\Delta P_d\right) \tag{5.18}$$

$$\Delta f = \left(I + 2\pi s^{-1} G_{gen} T\right)^{-1} \left[G_{gen} G_g G_t \left(-U\right) + G_{gen} \left(-\Delta P_d\right)\right] \tag{5.19}$$

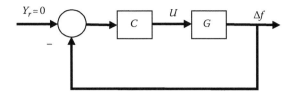

**FIGURE 5.5**
Unity feedback representation of a multi-area LFC system.

Comparing (5.14) and (5.19) when $\Delta P_d = 0$, we get

$$\overline{G} = \left(I + 2\pi s^{-1}G_{gen}T\right)^{-1}G_{gen}G_gG_t \qquad (5.20)$$

The control signal $U$ is given by

$$U = -C\left(2\pi s^{-1}T + B\right)\Delta f \qquad (5.21)$$

A unity feedback block diagram of (5.20) and (5.21) is shown in Figure 5.5, where

$$G = \left(2\pi s^{-1}T + B\right)\left(I + 2\pi s^{-1}G_{gen}T\right)^{-1}G_{gen}G_gG_t \qquad (5.22)$$

The transfer function matrix $M$ of the closed-loop interconnected system can be written from Figure 5.5 as

$$M = GC\left(I + GC\right)^{-1} \qquad (5.23)$$

The open-loop transfer function matrix $G$ defined in (5.22) can be decomposed into a diagonal matrix $\tilde{G}$ and an off-diagonal matrix $\hat{G}$ defined by

$$\tilde{G} = \mathrm{diag}\left[G_{ii}\right] \qquad (5.24)$$

$$\hat{G} = G - \tilde{G} \qquad (5.25)$$

where $G_{ii}$ are the diagonal elements of the matrix $G$ that represent an isolated subsystem when the interconnection matrix $T = 0$.

## 5.2.1 Decentralized Control Design for Multivariable Systems

The individually designed controller $C_i$ forms a stable closed-loop for an isolated LFC subsystem. In other words, the decentralized controllers $C = \mathrm{diag}[C_i]$ guarantee the stability of the closed-loop of the diagonal transfer function matrix $\tilde{G} = \mathrm{diag}\left[G_{ii}\right]$ shown in Figure 5.6.

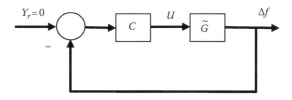

**FIGURE 5.6**
Closed-loop of diagonal transfer function matrix.

However, since the open-loop transfer function matrix $G$ is not diagonal, that is, $\hat{G} \neq 0$, the following question must be resolved. If the decentralized controllers are designed to form stable closed-loop diagonal systems, then what are the conditions that assure the stability of the closed-loop interconnected system given in Figure 5.5? The answer to this question has been addressed using the interaction measures based on structured singular values [2]. The procedure to check the stability of the closed-loop interconnected system under the decentralized control is presented here. To this end, the diagonal closed-loop transfer function matrix $\tilde{M}$ is determined from Figure 5.6 as

$$\tilde{M} = \tilde{G}C(I + \tilde{G}C)^{-1} \qquad (5.26)$$

Since each controller results in a stable closed-loop local subsystem, the matrix $\tilde{M}$ is stable under the diagonal controller $C$. However, an interaction measure (IM) will be defined to ensure the stability of the interconnected closed-loop matrix $M$ (5.23). The interaction term or the relative error matrix $E$ defined by

$$E = (G - \tilde{G})\tilde{G}^{-1} = \hat{G}\tilde{G}^{-1} \qquad (5.27)$$

is used to define an upper bound for the maximum singular value of the matrix $\tilde{M}$ such that the interconnected closed-loop system $M$ is stable. In [2], the following theorem is proved.

**Theorem [2]:**

The interconnected closed-loop system $M$ is stable if the following conditions are satisfied:

i. The matrices $G$ and $\tilde{G}$ have the same number of right half-plane (RHP) poles.
ii. The matrix $\tilde{M}$ is stable.

iii. $\bar{\sigma}\left(\tilde{M}(j\omega)\right) < \mu^{-1}\left(E(j\omega)\right)$ for all values of $\omega$.

where $\mu(\cdot)$ represents the structured singular value (SSV) suggested by Doyle [3] and $\bar{\sigma}(.)$ is the maximum singular value. The IM is the shortest distance between the two frequency response curves $\bar{\sigma}\left(\tilde{M}(j\omega)\right)$ and $\mu^{-1}(E(j\omega))$ [4].

Two examples are presented to explain the procedure of designing decentralized LFC controllers for two-area and three-area systems.

**Example 5.1**

An LFC power system consists of two identical areas [4] having the parameters listed in Table 5.1. First, we design decentralized controllers $C_1$ and $C_2$ using the two-step procedure presented in Section 4.5. The plant transfer function of each area is evaluated as

$$G_1 = G_2 = \frac{120}{(20s+1)(0.3s+1)(0.08s+1)}$$

and its SOPTD model is determined using (4.59) as

$$G_{SOPTD} = \frac{17.64}{(s+2.94)(s+0.05)}e^{-0.04s}$$

This approximate model can be factorized as

$$G_{SOPTD-} = \frac{17.64}{(s+2.94)(s+0.05)}$$

and

$$G_{SOPTD+} = e^{-0.04s}$$

**TABLE 5.1**

Parameters of Two Identical Areas of an LFC System

| Parameter | Value |
|---|---|
| Turbine time constant | $T_{t1}=T_{t2}=0.3$ s |
| Governor time constant | $T_{g1}=T_{g2}=0.08$ s |
| Generator time constant | $T_{geni} = \dfrac{2H_i}{D_i} = 20\,\text{s},\, i=1,2$ |
| Generator gain | $K_{geni} = \dfrac{1}{D_i} = 120\,\text{Hz/pu MW},\, i=1,2$ |
| Synchronizing power coefficient | $T_{12}=0.545$ pu  MW |
| Frequency bias | $B_1=B_2=0.425$ pu MW/Hz |

Using (4.47) and assuming $r=1$ and $T_f=0.1$, the controllers $C(s)$ take the form

$$C_1(s) = C_2(s) = \frac{(s+2.94)(s+0.05)}{17.64\left((1+0.1s) - e^{-0.04s}\right)}$$

Replacing $e^{-0.04s} \simeq (1-0.04s)$, the foregoing controllers will be

$$C_1(s) = C_2(s) = \frac{0.4(s+2.94)(s+0.05)}{s}$$

which are in PID form. The open-loop transfer function matrix

$$G = \begin{bmatrix} G_{11} & G_{12} \\ G_{21} & G_{22} \end{bmatrix}$$

is determined from (5.22) as

$$G_{11} = G_{22} = \frac{106.25\left(s^2 + 8.107s + 20.95\right)}{\Delta}$$

and

$$G_{12} = G_{21} = \frac{-856.084(s - 2.5)}{\Delta}$$

where

$$\Delta = (s+12.5)(s+3.33)(s+0.05)\left(s^2 + 0.05s + 41.1\right)$$

It is obvious that $\tilde{G} = \begin{bmatrix} G_{11} & 0 \\ 0 & G_{22} \end{bmatrix}$ and $\hat{G} = \begin{bmatrix} 0 & G_{12} \\ G_{21} & 0 \end{bmatrix}$. The diagonal closed-loop transfer function matrix $\tilde{M}$ with no interconnection is found from (5.26) as

$$\tilde{M}_{11} = \tilde{M}_{22} = \frac{106.25s\left(s^2 + 8.107s + 20.95\right)}{(s+11.15)(s+3.393)(s+1.016)(s+0.05)\left(s^2 + 0.3228s + 68.92\right)}$$

Since $G$ and $\tilde{G}$ have no RHP poles and $\tilde{M}$ is stable, the closed-loop interconnected system $M$ is stable if condition (iii) holds. To check that, the singular values of $\tilde{M}(j\omega)$ and the SSV of $E(j\omega)$ are plotted as shown in Figure 5.7. It is clear that condition (iii) is satisfied with IM = 3 dB. The response of frequency deviation when a disturbance of 5% occurs in both areas is shown in Figure 5.8.

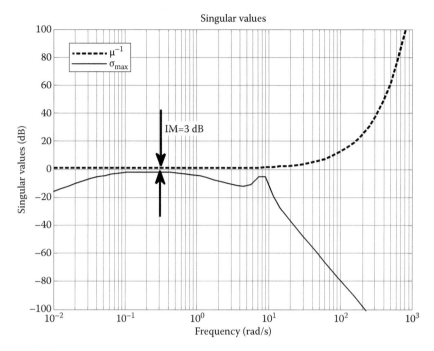

**FIGURE 5.7**

Singular value of $\tilde{M}(j\omega)$ and SSV of $E(j\omega)$.

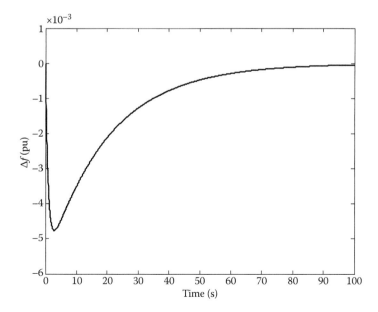

**FIGURE 5.8**

Frequency deviations in pu.

**Example 5.2**

The procedure to design a decentralized LFC outlined previously is explained in this example using a three-area LFC power system shown in Figure 5.9. The generating units in area 1, area 2, and area 3 are of non-reheat type, reheat type, and hydro type, respectively. The parameters of all units are given in Example 4.5. The governor droop constants are $R_1 = R_2 = R_3 = 0.05$. The synchronizing power coefficients are given as $T_{12} = T_{21} = T_{13} = T_{31} = T_{23} = T_{32} = 0.016$ and the bias factors are $B_1 = D_1 + (1/R_1) = 20.5$, $B_2 = D_2 + (1/R_2) = 20.25$, and $B_1 = D_1 + (1/R_1) = 21$. Three decentralized PI controllers are designed independently for each area.

Transfer functions of each isolated area and its FOPTD models have been determined in Example 4.5 and listed in Table 5.2.

Using the FOPTD model of each area, a PI controller is designed using equations (4.69) and (4.70), which are repeated here:

$$C(s) = \frac{1 + \tau_1 s}{s(T_f + \theta)} \tag{5.28}$$

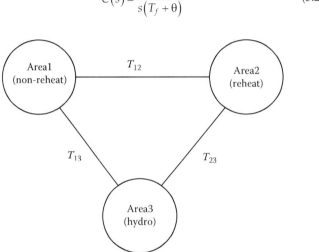

**FIGURE 5.9**
Three-area LFC power system.

**TABLE 5.2**

FOPDT Models of Each Isolated Area of a Three-Area LFC System

| Plant | FOPTD |
|---|---|
| $G_{non}(s) = \dfrac{2}{(1+24s)(1+0.3s)(1+0.2s)}$ | $\hat{G}_{area1} = \dfrac{2}{1+24.15s} e^{-0.35s}$ |
| $G_r(s) = \dfrac{4(1+2.1s)}{(1+40s)(1+7s)(1+0.3s)(1+0.25s)}$ | $\hat{G}_{area2}(s) = \dfrac{4}{1+42.45s} e^{-3s}$ |
| $G_h(s) = \dfrac{(1-2s)}{(1+10s)(1+0.5s)(1+s)}$ | $\hat{G}_{area3}(s) = \dfrac{1}{1+10.5s} e^{-3s}$ |

**TABLE 5.3**

Decentralized PI Controllers

| Area | 1 | 2 | 3 |
|------|---|---|---|
| PI controller | $C_1 = 17.88\left(1 + \dfrac{1}{24.15s}\right)$ | $C_2 = 12.13\left(1 + \dfrac{1}{42.45s}\right)$ | $C_3 = 1.31\left(1 + \dfrac{1}{10.5s}\right)$ |

$$G_s(s) = K_{c1}\left(1 + \frac{1}{T_{i1}s}\right) \tag{5.29}$$

with $K_{c1} = \tau_1/(T_f + \theta)$ and $T_{i1} = \tau_1$. Table 5.3 presents the decentralized PI controllers for the three areas.

The response of frequency deviation for the interconnected areas is shown in Figure 5.10 when load disturbances of 0.2, 0.1, and 0.05 pu take place at area 1, area 2, and area 3, respectively. The mechanical power is depicted in Figure 5.11.

The elements of the open-loop transfer function matrix $G = \begin{bmatrix} G_{11} & G_{12} & G_{13} \\ G_{21} & G_{22} & G_{23} \\ G_{31} & G_{32} & G_{33} \end{bmatrix}$ are found as

$$G_{11} = \frac{1.39\left(s^2 + 0.068s + 0.01\right)\left(s^2 + 0.06s + 0.03\right)}{\Delta_1}$$

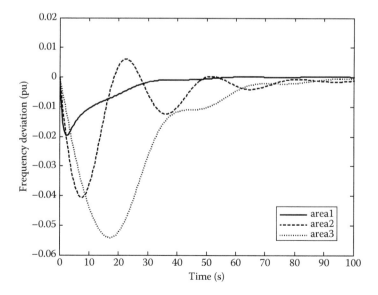

**FIGURE 5.10**
Frequency deviations in pu.

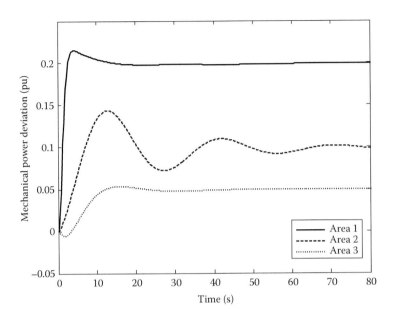

**FIGURE 5.11**
Mechanical power deviations in pu.

$$G_{12} = \frac{0.001(s+0.48)\left(s^2+0.1s+0.03\right)}{\Delta_1}$$

$$G_{13} = \frac{-0.003(s-0.5)\left(s^2+0.02s+0.03\right)}{\Delta_1}$$

$$G_{21} = \frac{0.01\left(s^2+0.1s+0.03\right)}{\Delta_2}$$

$$G_{22} = \frac{0.12(s+0.4762)(s^2+0.07s+0.01)(s^2+0.07s+0.03)}{\Delta_2}$$

$$G_{23} = \frac{-0.004(s-0.5)\left(s^2+0.04s+0.025\right)}{\Delta_2}$$

$$G_{31} = \frac{0.01\left(s^2+0.025s+0.03\right)}{\Delta_3}$$

$$G_{32} = \frac{0.001(s+0.48)\left(s^2+0.04s+0.025\right)}{\Delta_3}$$

and

$$G_{33} = \frac{-0.4(s-0.5)\left(s^2+0.03s+0.01\right)\left(s^2+0.03s+0.028\right)}{\Delta_3}$$

where

$$\Delta_1 = (s+5)(s+3.3)\Delta_4$$
$$\Delta_2 = (s+4)(s+0.14)\Delta_4$$
$$\Delta_3 = (s+2)(s+1)\Delta_4$$
$$\Delta_4 = (s+0.06)(s^2+0.08s+0.03)(s^2+0.03s+0.03)$$

Moreover, the elements of the diagonal closed-loop matrix $\tilde{M} = \text{diag}\left[\tilde{M}_{11}, \tilde{M}_{22}, \tilde{M}_{33}\right]$ are determined as

$$\tilde{M}_{11} = \frac{1.3889s\left(s^2+0.06752s+0.01083\right)}{\left(s+6.317\right)\left(s+0.04019\right)\left(s^2+0.0805s+0.01116\right)}$$

$$\tilde{M}_{22} = \frac{0.12s\left(s+0.4762\right)\left(s^2+0.0686s+0.009545\right)}{\left(s+3.639\right)\left(s+0.02167\right)\left(s^2+0.08717s+0.009601\right)\left(s^2+0.4898s+0.2061\right)}$$

$$\tilde{M}_{33} = \frac{-0.4s\left(s-0.5\right)\left(s^2+0.03496s+0.009135\right)}{\left(s+2.439\right)\left(s+0.3504\right)\left(s^2+0.07289s+0.003805\right)\left(s^2+0.2719s+0.07014\right)}$$

It is easy to verify that $\tilde{G} = \begin{bmatrix} G_{11} & 0 & 0 \\ 0 & G_{22} & 0 \\ 0 & 0 & G_{33} \end{bmatrix}$ and $G = \begin{bmatrix} G_{11} & G_{12} & G_{13} \\ G_{21} & G_{22} & G_{23} \\ G_{31} & G_{32} & G_{33} \end{bmatrix}$

have no RHP poles and $\tilde{M}$ is stable. Therefore, the interconnected closed-loop transfer function matrix $M$ is stable since condition (iii) is satisfied with interaction measure IM = 6.8 dB, as shown in Figure 5.12.

## 5.3 Decentralized PID Controllers for LFC with Time Delay

The load frequency control scheme requires transmission of the measured ACE signals from remote terminal units (RTUs) to the control center and transmission of control signals from the control center to the generating units. In conventional LFC, signals are transmitted via dedicated communication channels, which is the responsibility of the large utilities. In case of channel failure, a backup is provided by voice communication via telephone lines. The use of a dedicated communication channel in an LFC system introduces constant time delay.

An effective power system market highly needs an open communication infrastructure to support the increasing decentralized property of control services such as LFC [5–9]. The open communication infrastructure will also allow a bilateral contract for the provision of load following and

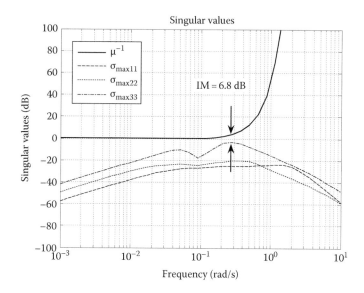

**FIGURE 5.12**
Singular values of $\tilde{H}(j\omega)$ and SSV of $E(j\omega)$ for a three-area system.

third-party LFC. With the introduction of open communication channels, both constant and time-varying delays are introduced in the LFC scheme [5]. In this section, the design of PID LFC controllers in the presence of constant time delay is considered.

### 5.3.1 LFC with Time Delay

The conventional LFC scheme developed in Chapter 1 is modified to include time delay into the control loop of both single-area and multi-area systems. In an open communication system, delays can arise during (1) transmission of control signals from the control center to the individual units and (2) transmission of ACE from RTUs to the control center. The model considered here aggregates all such delays into a single delay from the control center [10].

Assume that a signal $z(t)$ experiences a fixed time delay $\theta_o$, then the delayed signal is represented as $z(t-\theta_o)$. The Laplace transform of the delayed signal is given by

$$\mathcal{L}\left(z\left(t-\theta_o\right)\right) = e^{-\theta_o s}Z\left(s\right) \tag{5.30}$$

where $\mathcal{L}(z(t)) = Z(s)$. Therefore, the time delay can be represented in block diagram form, as shown in Figure 5.13.

**FIGURE 5.13**
Representation of a time-delayed signal.

### 5.3.1.1 Single-Area LFC with Time Delay

The time delay associated with ACE and control input signals are lumped into a single delay $\theta_o$. A block diagram model of a single-area LFC scheme with time delay is given in Figure 5.14. The turbine transfer function $G_t$ might be of non-reheat, reheat, or hydro type.

The open-loop transfer function is given by

$$G = G_g G_t G_{gen} e^{-\theta_o s} \tag{5.31}$$

The rules given by Equations 4.56 through 4.59 can be used to determine an approximate FOPTD or SOPTD model of (5.31). Then, the design of PI or PID controllers is straightforward, as explained in the next two examples.

**Example 5.3**

Consider a single area with non-reheat turbine having the following parameters: $T_t = 0.3$, $T_g = 0.1$, $R = 0.05$, $D = 1$, and $H = 5$. A constant time delay of 20 s is assumed [10]. The open loop transfer function is calculated using (5.31) as

$$G(s) = \frac{1}{(1 + \tau_{1o}s)(1 + \tau_{2o}s)(1 + \tau_{3o}s)} e^{-\theta_o s}$$

with $\tau_{1o} = 10$, $\tau_{2o} = 0.3$, $\tau_{3o} = 0.1$, and $\theta_o = 20$. This transfer function can be approximated by a FOPTD model $\hat{G}(s) = (1/(1 + \tau_1 s)) e^{-\theta s}$

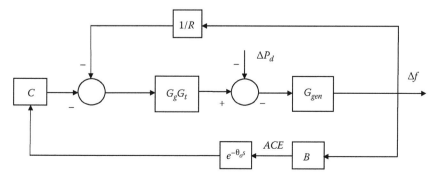

**FIGURE 5.14**
LFC scheme of a single-area with time delay.

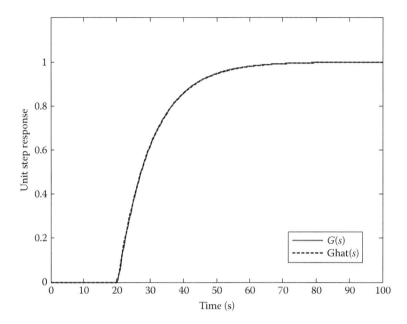

**FIGURE 5.15**
Unit step response of the open-loop plant and its FOPTD model.

where $\tau_1 = \tau_{1o} + 0.5\tau_{2o} = 10.15$ and $\theta = \theta_o + 0.5\tau_{2o} + \tau_{3o} = 20.25$. To validate the approximation, the unit step response of both the plant and its FOPTD model are shown in Figure 5.15. Assuming $T_f = 20$, the PI controller parameters are calculated using Equation 4.71 as $K_{c1} = \tau_1/(T_f + \theta) = 10.15/(20 + 20.25) = 0.252$ and $T_{i1} = \tau_1 = 10.15$. The frequency and mechanical power deviations when a load disturbance of 10% is applied at time $t = 50$ s are shown in Figures 5.16 and 5.17, respectively.

**Example 5.4**

The three-area power system presented in [10] in the context of robust LFC with communication delay is used in this example to design decentralized PI controllers. The system data are given in Table 5.4. The time delay for each area is assumed to have the following ranges $\theta_{o1} \in [0\ 10]$, $\theta_{o2} \in [0\ 5]$, and $\theta_{o3} \in [0\ 5]$. The FOPTD model of each area is determined using the maximum value of time delay. Table 5.5 shows the open-loop transfer function of each area and its FOPTD model.

The PI controllers designed for each isolated area based on the FOPTD models are determined as $C_{area1} = 0.667(1 + (1/20.15s))$, $C_{area2} = 0.032(1 + (1/8.2s))$, and $C_{area3} = 0.027(1 + (1/6.84s))$. These controllers are then simulated for the interconnected three-area system. A load disturbance of 10% at $t = 4$ s, 20% at $t = 2$ s, and 30% at $t = 2$ s are assumed at area 1, area 2, and area 3, respectively. To check the robustness of the controllers against the variation of time delay, simulation is carried out for three different cases of time delay, as shown in Table 5.6.

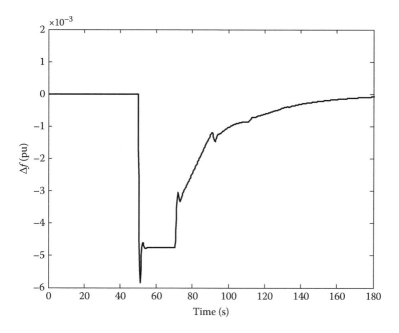

**FIGURE 5.16**
Frequency deviation due to 10% load disturbance.

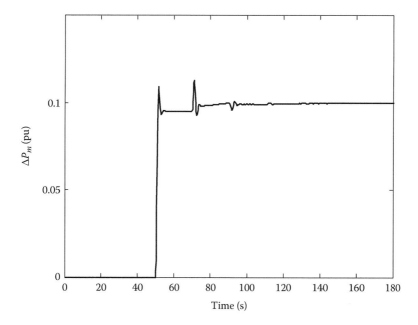

**FIGURE 5.17**
Mechanical power deviation due to 10% load disturbance.

**TABLE 5.4**

Parameters of Example 5.4

| Parameters | Units | Area 1 | Area 2 | Area 3 |
|---|---|---|---|---|
| $T_{geni}$ | s | 20 | 8 | 6.667 |
| $K_{geni}$ | pu Hz/pu MW | 2 | 0.667 | 0.556 |
| $T_{gi}$ | s | 0.1 | 0.4 | 0.35 |
| $T_{ti}$ | s | 0.3 | 0.17 | 0.2 |
| $R_i$ | pu Hz/pu MW | 0.05 | 0.05 | 0.05 |
| $T_{ij}$ | pu MW | $T_{12} = 0.4$ | $T_{21} = T_{12}$ | $T_{31} = T_{13}$ |
|  |  | $T_{13} = 0.4$ | $T_{23} = 0.4$ | $T_{32} = T_{23}$ |
| $B_i = D_i + \dfrac{1}{R_i}$ | pu MW/puHz | 20.5 | 81.5 (4 units) | 81.8 (4 units) |

**TABLE 5.5**

FOPTD of the Open-Loop Transfer Function of Each Area of Example 5.4

| Area | FOPTD |
|---|---|
| $G_{area1}(s) = \dfrac{2}{(1+20s)(1+0.3s)(1+0.1s)} e^{-10s}$ | $\hat{G}_{area1} = \dfrac{2}{1+20.15s} e^{-10.25s}$ |
| $G_{area2}(s) = \dfrac{0.667}{(1+8s)(1+0.4s)(1+0.17s)} e^{-5s}$ | $\hat{G}_{area2} = \dfrac{0.667}{1+8.2s} e^{-5.37s}$ |
| $G_{area3}(s) = \dfrac{0.556}{(1+6.667s)(1+0.35s)(1+0.2s)} e^{-5s}$ | $\hat{G}_{area3} = \dfrac{0.556}{1+6.842s} e^{-5.375s}$ |

**TABLE 5.6**

Three Cases of Time Delay

| Case | 1 (No Delay) | 2 (Maximum Delay) | 3 (200% of Maximum Delay) |
|---|---|---|---|
| Time delay | $\theta_{o1} = 0$ | $\theta_{o1} = 10$ | $\theta_{o1} = 20$ |
|  | $\theta_{o2} = 0$ | $\theta_{o2} = 5$ | $\theta_{o2} = 10$ |
|  | $\theta_{o3} = 0$ | $\theta_{o3} = 5$ | $\theta_{o3} = 10$ |

Simulation results of frequency deviation, area tie-line power, and mechanical power deviation for case 1 are shown in Figure 5.18. The second case representing the maximum delay is simulated for the same load disturbances and the results are given in Figure 5.19. Simulation results of case 3 are presented in Figure 5.20.

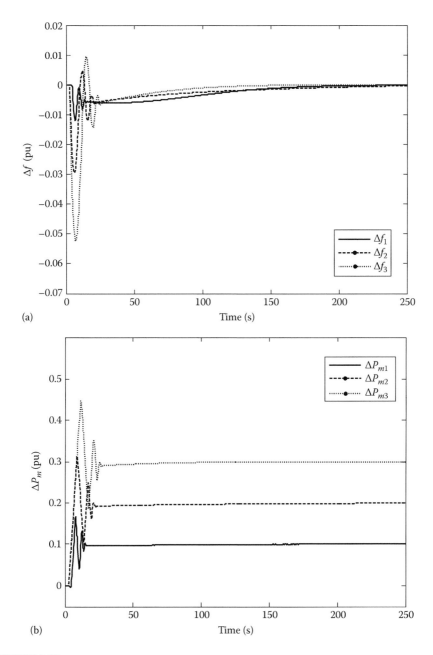

(a)

(b)

**FIGURE 5.18**
Simulation results of a three-area system equipped with decentralized PI controllers (case 1):
(a) frequency deviation, (b) mechanical power deviation. (*Continued*)

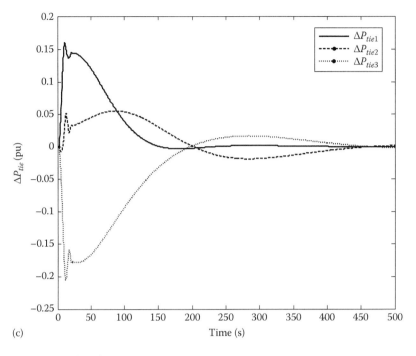

(c)

**FIGURE 5.18 (*Continued*)**
Simulation results of a three-area system equipped with decentralized PI controllers (case 1): (c) area-tie-line power.

The SOPTD models given in Table 5.7 are used to design PID controllers for each isolated area. The designed controllers are determined as

$$C_{area1} = \frac{0.06125s^2 + 0.1781s + 0.00875}{7s^2 + 0.8759s}$$

$$C_{area2} = \frac{0.03613s^2 + 0.07901s + 0.009312}{3.88s^2 + 0.342s}$$

$$C_{area3} = \frac{0.02592s^2 + 0.06149s + 0.00864}{3s^2 + 0.2645s}$$

The interconnected three-area system is simulated in the presence of these decentralized PID controllers. The load disturbance is assumed to be 10%, 20%, and 30% in area 1, area 2, and area 3, respectively. Frequency deviations are shown in Figure 5.21 for the case of maximum time delay.

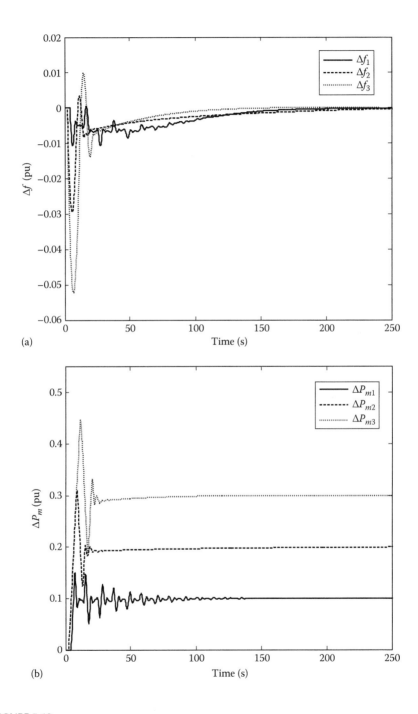

(a)

(b)

**FIGURE 5.19**
Simulation results of a three-area system equipped with decentralized PI controllers (case 2):
(a) frequency deviation, (b) mechanical power deviation. (*Continued*)

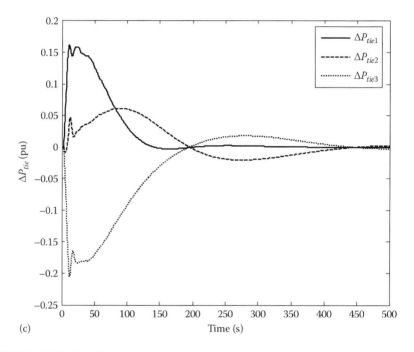

(c)

**FIGURE 5.19 (*Continued*)**
Simulation results of a three-area system equipped with decentralized PI controllers (case 2): (c)
area-tie-line power.

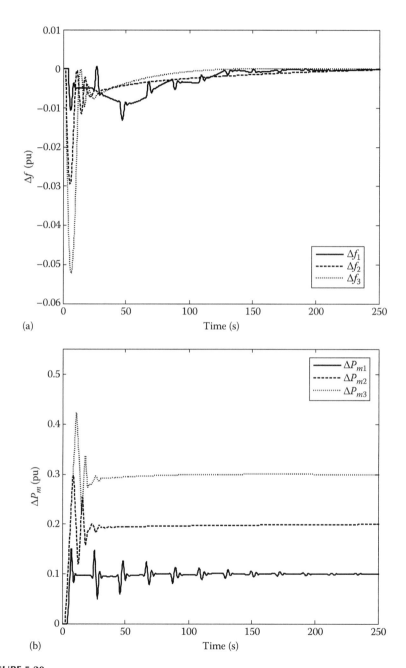

**FIGURE 5.20**
Simulation results of a three-area system equipped with decentralized PI controllers (case 3): (a) frequency deviation, (b) mechanical power deviation. *(Continued)*

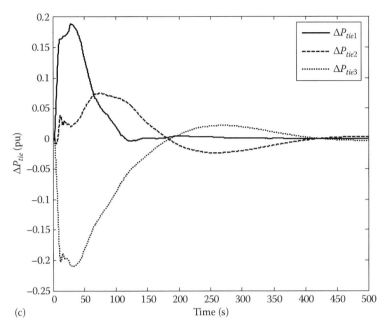

(c)

**FIGURE 5.20 (*Continued*)**
Simulation results of a three-area system equipped with decentralized PI controllers (case 3):
(c) area-tie-line power.

**TABLE 5.7**

SOPTD of the Open-Loop Transfer Function of Each Area of Example 5.4

| Area | SOPTD |
|---|---|
| $G_{area1}(s) = \dfrac{2}{(1+20s)(1+0.3s)(1+0.1s)} e^{-10s}$ | $\hat{G}_{area1} = \dfrac{2}{(1+20s)(1+0.35s)} e^{-10.05s}$ |
| $G_{area2}(s) = \dfrac{0.667}{(1+8s)(1+0.4s)(1+0.17s)} e^{-5s}$ | $\hat{G}_{area2} = \dfrac{0.667}{(1+8s)(1+0.485s)} e^{-5.08s}$ |
| $G_{area3}(s) = \dfrac{0.556}{(1+6.667s)(1+0.35s)(1+0.2s)} e^{-5s}$ | $\hat{G}_{area3} = \dfrac{0.556}{(1+6.667s)(1+0.45s)} e^{-5.1s}$ |

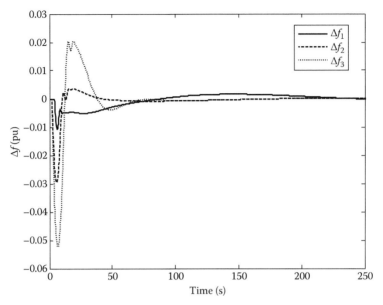

**FIGURE 5.21**
Frequency deviations of a three-area system equipped with decentralized PID controllers for maximum time delay.

# References

1. W. Tan, Unified tuning of PID load frequency controller for power systems via IMC, *IEEE Trans. Power Syst.*, 25(1), 341–350, 2010.
2. P. Grosdidier and M. Morari, Interaction measures for system under decentralized control, *Automatica*, 22, 341–350, 1986.
3. J. Doyle, Analysis of feedback systems with structured uncertainties, *IEE Proc. Pt. D*, 129(6), 242–250, November 1982.
4. T. C. Yang, T. Ding and H. Yu, Decentralized power system load frequency control beyond the limit of diagonal dominance, *Electr. Power Energy Syst.*, 24, 173–184, 2002.
5. S. Bhowmik, K. Tomsovic, and A. Bose, Communication models for third party load frequency control, *IEEE Trans. Power Syst.*, 19(1), 543–548, February 2004.
6. H. Shayeghi, H. A. Shayanfar, and A. Jalili, Multi-stage fuzzy PID power system automatic generation controller in deregulated environments, *Energy Convers. Manage.*, 47(18), 2829–2845, November 2006.
7. H. Bevrani, *Robust Power System Frequency Control*, Springer, New York, 2009.
8. V. Donde, M. A. Pai, and I. A. Hiskens, Simulation and optimization of an AGC after deregulation, *IEEE Trans. Power Syst.*, 16(3), 481–489, August 2001.

9. L. Jiang, W. Yao, Q. H. Wu, J. Y. Wen, and S. J. Cheng, Delay-dependent stability for load frequency control with constant and time-varying delays, *IEEE Trans. Power Syst.*, 27(2), 932–941, May 2012.
10. X. Yu and K. Tomsovic, Application of linear matrix inequalities for load frequency control with communication delays, *IEEE Trans. Power Syst.*, 19(3), 1508–1515, August 2004.

# Part II

# Fuzzy Logic Control Approaches

# 6

# *Fuzzy Systems and Functions Approximation*

## 6.1 Introduction

Control engineering applications rely on the information available from two basic sources: the numerical data available from sensors and the descriptive (linguistic) information available from human experts. The numerical information can be used in conventional control systems. However, the linguistic information cannot be used directly in the conventional controller design. In order to incorporate this type of information in control systems, a methodology is needed to represent this information in a quantified way. Different fuzzy systems have been extensively used to represent the linguistic information [1]. In this chapter, a brief description of different types of fuzzy systems is presented. The function approximation capability and fuzzy basis functions (FBF) representation of fuzzy systems are explained.

## 6.2 Types of Fuzzy Systems

Fuzzy systems are categorized into three classes: pure fuzzy systems, Takagi–Sugeno fuzzy systems, and fuzzifier–defuzzifier fuzzy systems.

### 6.2.1 Pure Fuzzy Systems

This type of fuzzy systems consists of two main parts, fuzzy rule base and fuzzy inference mechanism, as depicted in Figure 6.1. The fuzzy rule base consists of a collection of $L$ linguistic fuzzy rules in the form of IF-THEN rules. The fuzzy inference mechanism is based on fuzzy logic principles, a mapping

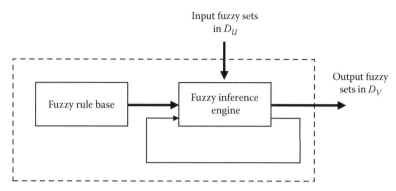

**FIGURE 6.1**
Basic components of a pure fuzzy system.

from fuzzy sets in the input universe of discourse $D_U \subset \Re^n$ to fuzzy sets in the output universe of discourse $D_V \subset \Re$, where $\Re^n$ represents the Euclidean space of dimension $n$. The $i$th fuzzy rule $R^i, i=1, \ldots, L$ takes the form

$$R^i : IF\ x_1\ is\ F_1^i\ \ and\ \ x_n\ is\ F_n^i\ THEN\ y\ is\ G^i \tag{6.1}$$

where

$F_j^i$ and $G^i$ are fuzzy sets
$x = [x_1 \cdots x_n] \in D_U$ and $y \in D_V$ are the input and output linguistic variables, respectively

### 6.2.2 Takagi–Sugeno Fuzzy Systems

In this type, the IF-THEN rules take the form [2]

$$R^i : IF\ x_1\ is\ F_1^i\ \ and\ \ x_n\ is\ F_n^i\ THEN\ y^i = a_0^i + a_1^i x_1 + \cdots + a_n^i x_n \tag{6.2}$$

The IF part of (6.2) is fuzzy and the THEN part is crisp. The output of (6.2) is a linear combination of input variables, where $a_0^i, a_1^i, \ldots, a_n^i$ are real constants. The inference mechanism used to find the aggregated output $y(x)$ is the weighted average of the form

$$y(x) = \frac{\sum_{i=1}^{L} w^i y^i}{\sum_{i=1}^{L} w^i} \tag{6.3}$$

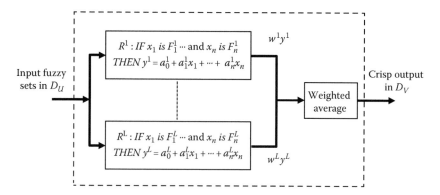

**FIGURE 6.2**
Basic components of a Takagi–Sugeno fuzzy system.

where the weight $w^i$ is calculated using

$$w^i = \prod_{l=1}^{n} \mu_{F_l^i}(x_l) \tag{6.4}$$

with $\mu_{F_l^i}(x_l)$ is defined as the membership function of the variable $x_l$ for the fuzzy set $F_l^i$. The basic configuration of the Takagi–Sugeno fuzzy system is shown in Figure 6.2.

### 6.2.3 Fuzzy Systems with a Fuzzifier and a Defuzzifier

The pure fuzzy system shown in Figure 6.1 is modified in a way such that it can be used directly in engineering applications where the inputs and outputs are real variables. The modified version includes a fuzzifier and a defuzzifier. The fuzzifier is a mapping from crisp points in $D_U$ into fuzzy sets in $D_U$ while the defuzzifier is another mapping from the output fuzzy sets in $D_V$ into crisp values in $D_V$. The basic components of a fuzzy system with a fuzzifier and a defuzzifier are depicted in Figure 6.3. This fuzzy system has been used mainly as a controller [3] and hence it is often called the fuzzy logic controller. We consider here and in the following sections this type of fuzzy system.

## 6.3 Details of a Fuzzy System with a Fuzzifier and a Defuzzifier

In this section, we give the fundamental details of the fuzzy logic system shown in Figure 6.3.

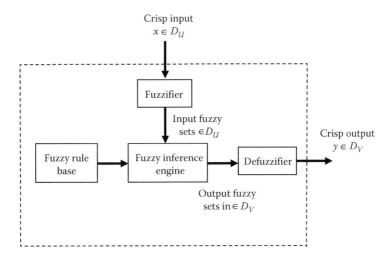

**FIGURE 6.3**
Basic components of a fuzzy system with a fuzzifier and a defuzzifier.

### 6.3.1 Definitions

1. *Fuzzy set*: A fuzzy set $F$ in a universe of discourse $D_U$ is characterized by a membership function $\mu_F:D_U \rightarrow [0,1]$ with $\mu_F(v)$ representing the grade of membership of $v \in D_U$ in the fuzzy set $F$. Figure 6.4 illustrates the membership functions of five fuzzy sets, Negative big (NB), Negative small (NS), Zero (Z), Positive small (PS), and Positive big (PB) for the variable $v$. The universe of discourse $D_U$ contains all possible values $v \in [v_{min}, v_{max}]$. The degree of membership ranges from zero to unity.

2. *Support*: The support of the fuzzy set $F$ is the crisp set of all points $v \in D_U$ so that $\mu_F(v) > 0$.

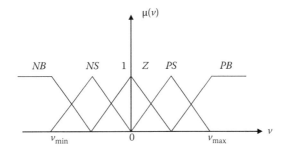

**FIGURE 6.4**
Membership functions of five fuzzy sets, NB, N, Z, PS, and PB for the variable $v$.

3. *Fuzzy singleton*: If the support of a fuzzy set is a single point in $D_U$ at which $\mu_F(v) = 1$, then $F$ is called a fuzzy singleton.

4. *Fuzzy implication*: For two fuzzy sets $A$ and $B$ in two universe of discourses $D_1$ and $D_2$, a fuzzy implication $A \to B$ is a special type of fuzzy relation in $D_1 \times D_2$. In fact, there are different types of fuzzy implications as presented in [1]. The choice of fuzzy implication is mainly based on the application and the computational burden. For the development of adaptive fuzzy control, two types of fuzzy implications have been used, the minimum inference rule and the product-inference rule.

## 6.3.2 Singleton Fuzzifier

Fuzzification is a mapping from a crisp point in the universe of discourse $D_U$ into a fuzzy set $F'$ in $D_U$. Three types of fuzzifiers are available in the literature: the singleton, the Gaussian, and the triangular fuzzifiers. The singleton fuzzifier simplifies the computation involved in the fuzzy inference engine and hence results in a simple fuzzy logic system. This type of fuzzifier is considered here. The singleton fuzzifier maps a real-valued point $x' \in D_U$ into a fuzzy singleton $F'$ with support $x = [x_1 \cdots x_n]^T \in D_U$, which has a membership value $\mu_F'(x') = 1$ for $x' = x$ and $\mu_F'(x) = 0$ for all other $x' \in D_U$ with $x' \neq x$.

## 6.3.3 Fuzzy Inference Mechanism

In the fuzzy inference mechanism, the IF-THEN rules are combined using fuzzy logic principles to produce a mapping from fuzzy sets in $D_U$ to fuzzy sets in $D_V$. Note that the IF-THEN rules in the form (6.1) can be interpreted as a fuzzy implication $A \to B$ in $D_U \times D_V$, where $A = F_1^i \times \cdots \times F_n^i$ and $B = G^i$. For $x = [x_1 \cdots x_n] \in D_U$ and $y \in D_V$, the min-inference and the product-inference rules as fuzzy inference mechanisms are given by (6.5) and (6.6), respectively:

$$\mu_{A \to B}(x,y) = \min\left(\mu_A(x), \mu_B(y)\right) \tag{6.5}$$

$$\mu_{A \to B}(x,y) = \mu_A(x)\mu_B(y) \tag{6.6}$$

where $\mu_{F_1^i \times \cdots \times F_n^i}(x)$ is calculated using the min-operation (6.7) or the product operation (6.8)

$$\mu_{F_1^i \times \cdots \times F_n^i}(x) = \min\left(\mu_{F_1^i}(x), \ldots, \mu_{F_n^i}(x)\right) \tag{6.7}$$

$$\mu_{F_1^i \times \cdots \times F_n^i}(x) = \mu_{F_1^i}(x) \cdots \mu_{F_n^i}(x) \tag{6.8}$$

### 6.3.4 Defuzzification

Mapping from fuzzy sets in $D_V$ to a crisp output $y \in D_V$ is termed as defuzzification. There are three possible defuzzifiers: the maximum output, the center average, and the modified center average. The center average defuzzifier has been used extensively in fuzzy approximation and control system applications [4]. This type of defuzzifier is given by

$$f(x) = \frac{\sum_{i=1}^{L} \bar{y}^i \left( \mu_{B^i} \left( \bar{y}^i \right) \right)}{\sum_{i=1}^{L} \mu_{B^i} \left( \bar{y}^i \right)} \tag{6.9}$$

where $\bar{y}^i$ is the point in $D_V$ at which $\mu_{G^i}(y)$ reaches its maximum value and $\mu_{B^i}(y)$ is given by

$$\mu_{B^i}\left( y \right) = \sup_{x \in D_U} \left( \mu_{F_1^i \times \cdots \times F_n^i \to G^l} \left( x, y \right) \mu_{A'} \left( x \right) \right) \tag{6.10}$$

### 6.3.5 Well-Known Fuzzy System

A well-known fuzzy system used in function approximation and control applications is the one that has a center average defuzzifier, product-inference rule, and a singleton fuzzifier. The form of such type of fuzzy system can be obtained as follows. Using (6.6) and (6.8) in (6.10), we get

$$\mu_{B^i}\left( \bar{y}^i \right) = \sup_{x' \in D_U} \left( \prod_{k=1}^{n} \mu_{F_k^i}(x_k') \mu_{G^i} \left( \bar{y}^i \right) \mu_{A'} \left( x' \right) \right) \tag{6.11}$$

where sup(.) stands for the supremum of the indicated function. For the singleton fuzzifier, $\mu_{A'}(x') = 1$ for $x' = x$ and $\mu_{A'}(x') = 0$ for all other $x' \in D_U$ and the sup in (6.11) is achieved at $x' = x$. Assuming the maximum of $\mu_{G^i}(y^i)$ is 1 and achieved at $\bar{y}^i$, that is to say, $\mu_{G^i}\left( \bar{y}^i \right) = 1$, then (6.11) is reduced to

$$\mu_{B^i}\left( \bar{y}^i \right) = \prod_{k=1}^{n} \mu_{F_k^i}(x_k') \tag{6.12}$$

Substituting (6.12) into (6.9), we get

$$f(x) = \frac{\sum_{i=1}^{L} \bar{y}^i \left( \prod_{k=1}^{n} \mu_{F_k^i}\left( x_k \right) \right)}{\sum_{i=1}^{L} \left( \prod_{k=1}^{n} \mu_{F_k^i}\left( x_k \right) \right)} \tag{6.13}$$

The shape of the membership functions can be triangular, trapezoidal, or Gaussian. A general class of membership functions is the pseudo-trapezoid membership function- which represents the triangular, trapezoidal, and Gaussian membership functions as special cases. Consider a fuzzy set $F$ defined in a universe of discourse $[\alpha, \delta] \in \mathfrak{R}$. The pseudo-trapezoid membership function for the fuzzy set $F$ is defined as

$$\mu_F\left(x; \alpha, \beta, \gamma, \delta\right) = \begin{cases} P(x), x \in [\alpha, \beta) \\ 1, x \in [\beta, \gamma] \\ N(x), x \in (\gamma, \delta] \\ 0, x \in \mathfrak{R} - (\alpha, \delta) \end{cases} \tag{6.14}$$

where $\alpha \leq \beta \leq \gamma \leq \delta, 0 \leq P(x) \leq 1$ and $0 \leq N(x) \leq 1$. The function $P(x)$ has positive slope in $[\alpha, \beta)$ while $N(x)$ has negative slope in $(\gamma, \delta]$. A triangular membership function is obtained from (6.14) by letting $\beta = \gamma$ and $P(x)$ and $N(x)$ take the form

$$P(x) = \frac{x - \alpha}{\beta - \alpha} \tag{6.15}$$

$$N(x) = \frac{\delta - x}{\delta - \gamma} \tag{6.16}$$

If $\beta \neq \gamma$ and $P(x)$ and $N(x)$ are determined from (6.15) and (6.16), then (6.14) becomes a trapezoidal membership function. If $\alpha = -\infty, \beta = \gamma = \overline{x}, \delta = \infty$ and $P(x) = N(x) = e^{-((x - \overline{x})/\delta)^2}$, then a Gaussian membership function of center $\overline{x}$ and width $\delta$ is obtained. The Gaussian membership function is the most commonly used function and it has the following form:

$$\mu_{F_k^i}\left(x_k\right) = e^{-\left(\frac{x_k - \overline{x}_k^i}{\sigma_k^i}\right)^2} \tag{6.17}$$

where $\overline{x}_k^i$ and $\sigma_k^i$ are the center and width of the $i$th fuzzy set $F_k^i$. The fuzzy system (6.13) with Gaussian membership function becomes

$$f(x) = \frac{\sum_{i=1}^{L} \overline{y}^i \left( \prod_{k=1}^{n} e^{-\left((x_k - \overline{x}_k^i)/\sigma_k^i\right)^2} \right)}{\sum_{i=1}^{L} \left( \prod_{k=1}^{n} e^{-\left((x_k - \overline{x}_k^i)/\sigma_k^i\right)^2} \right)} \tag{6.18}$$

## 6.4 Universal Functions Approximation Using Fuzzy Systems

The fuzzy logic system given by (6.13) is capable of approximating any nonlinear function defined in $D_U$ to any degree of accuracy if $D_U$ is compact. The universal approximation theorem states that for any given real continuous function $g(x)$ defined on a compact set $D_U \in \mathfrak{R}^n$ and arbitrary constant $\varepsilon > 0$, there exists a fuzzy system $f(x)$ in the form of (6.13) such that

$$\sup_{x \in D_U} |f(x) - g(x)| < \varepsilon \qquad (6.19)$$

The proof of this theorem is given in [4].

### 6.4.1 Design of a Fuzzy System

In this subsection, we are going to show how to construct a fuzzy system in the form (6.13) that approximates an unknown function $g(x)$. Without loss of generality, consider a nonlinear function $g(x_1, x_2)$ defined on the compact set $U = x_1 \times x_2 \subset \mathfrak{R}^2$, where $x_1 \in [a_1, b_1]$ and $x_2 \in [a_2, b_2]$. Suppose that $g(x_1, x_2)$ is available for any $(x_1, x_2) \in U$. It is required to design a two-input fuzzy system in the form of (6.13) with triangular membership function to uniformly approximate $g(x_1, x_2)$. The next steps are followed to construct the required fuzzy system. It is to be noted that the procedure can be applied for nonlinear function with $n$-variable and for any membership function of the general form (6.14).

*Step 1:* Define fuzzy sets $F_1^1, \ldots, F_1^{n_1}$ for $x_1$ in $[a_1, b_1]$ and $F_2^1, \ldots, F_2^{n_2}$ for $x_2$ in $[a_2, b_2]$, where $n_1$ and $n_2$ are the number of fuzzy sets for the inputs $x_1$ and $x_2$, respectively. The membership functions for these fuzzy sets are denoted by $\mu_{F_i^{n_i}}\left(x; \alpha_i^k, \beta_i^k, \delta_i^k\right)$ for $i = 1, 2$ and $k = 1, \ldots, n_i$ with $\alpha_i^1 = a_i$ and $\delta_i^1 = b_i$. For $x_1$, define $h_1^{1} = a_1, h_1^{m} = b_1, \Delta_1 = (b_1 - a_1)/(n_1 - 1)$ and $h_1^j = h_1^{j-1} + (j - 1)\Delta_1, j = 2, 3, \ldots, (n_1 - 1)$. Similarly, for $x_2$, define $h_2^1 = a_2, h_2^{n_2} = b_2, \Delta_2 = (b_2 - a_2)/(n_2 - 1)$ and $h_2^j = h_2^{j-1} + (j - 1)\Delta_2, j = 2, 3, \ldots, (n_2 - 1)$.

*Step 2:* Construct $m$ fuzzy IF-THEN rules of the form

$$R^{ik} : \text{IF } x_1 \text{ is } F_1^i \text{ and } x_2 \text{ is } F_2^k, \text{ THEN } y \text{ is } B^{ik} \qquad (6.20)$$

where $m = n_1 \times n_2, i = 1, \ldots, n_1, k = 1, \ldots, n_2$ and $B^{ik}$ is a fuzzy set whose center is $\bar{y}^{ik}$ chosen as

$$\bar{y}^{ik} = g\left(h_1^i, h_2^k\right) \qquad (6.21)$$

*Step 3:* Construct the fuzzy system $f(x_1, x_2)$ from the $m$ rules defined in (6.20) using product inference, singleton fuzzifier, and center average defuzzifier.

In this case, $f(x_1, x_2)$ takes the following form

$$f(x_1, x_2) = \frac{\sum_{i=1}^{n_1} \sum_{k=1}^{n_2} \overline{y}^{ik} \mu_{F_1^i}(x_1) \mu_{F_2^k}(x_2)}{\sum_{i=1}^{n_1} \sum_{k=1}^{n_2} \mu_{F_1^i}(x_1) \mu_{F_2^k}(x_2)} \qquad (6.22)$$

### Example 6.1

In this example, we illustrate the use of the preceding steps in constructing a fuzzy system to approximate the function $g(x) = \cos(x)$ defined on $U = [-4, 4]$. Seven fuzzy sets are chosen with triangular membership functions with $\Delta = 8/6$, $h^1 = -4$, $h^7 = 4$, and $h^j = h^{j-1} + (j-1)\Delta$, $j = 2, \ldots, 6$. These membership functions will be $\mu_{F^1}(x; -4, -4, -4+8/6)$, $\mu_{F^7}(x; 8/3, 4, 4)$, $\mu_{F^j}(x; h^{j-1}, h^j, h^{j+1})$, and $j = 2, \ldots, 6$. For example, the membership function $\mu_{F^1}(x; -4, -4, -4+8/6)$ has $\alpha = -4$, $\beta = -4$, and $\delta = -8/3$; then according to (6.16) its shape will be $\mu_{F^1}(x; -4, -4, -4+8/6) = (\delta - x)/(\delta - \beta) = -(3/4)x - 2$. The membership functions are shown in Figure 6.5.

The required fuzzy system is determined as

$$f_7(x) = \frac{\sum_{j=1}^{7} \cos(h^j) \mu_{F^j}(x; h^{j-1}, h^j, h^{j+1})}{\sum_{j=1}^{7} \mu_{F^j}(x; h^{j-1}, h^j, h^{j+1})} \qquad (6.23)$$

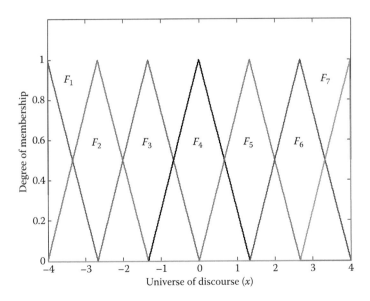

**FIGURE 6.5**
Seven membership functions of Example 6.1.

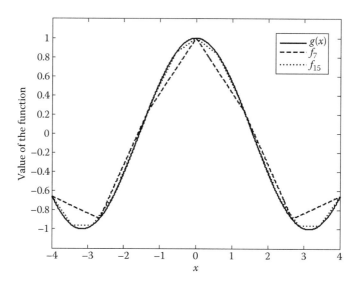

**FIGURE 6.6**
Function of Example 6.1 and its fuzzy approximation.

If 15 fuzzy sets are used, then $\Delta = 8/14$, $h^1 = -4, h^{15} = 4$ and $h^j = h^{j-1} + (j-1)\Delta$, $\mu_{F^1}(x; -4, -4, -4 + 8/14)$, $\mu_{F^7}(x; 48/14, 4, 4)$, $\mu_{F^j}(x; h^{j-1}, h^j, h^{j+1})$, and $j = 2, \ldots, 14$. For this case, the fuzzy system becomes

$$f_{15}(x) = \frac{\sum_{j=1}^{15} \cos\left(h^j\right)\mu_{F^j}\left(x; h^{j-1}, h^j, h^{j+1}\right)}{\sum_{j=1}^{15} \mu_{F^j}\left(x; h^{j-1}, h^j, h^{j+1}\right)} \qquad (6.24)$$

The function $g(x) = \cos(x)$ and its fuzzy approximations $f_7(x)$ and $f_{15}(x)$ are plotted in Figure 6.6.

The infinite norm of the fuzzy approximation error denoted by $\|f(x) - g(x)\|_\infty$ is calculated for both cases as $\|f_7(x) - g(x)\|_\infty = 0.21$ and $\|f_{15}(x) - g(x)\|_\infty = 0.04$. This means that, the fuzzy approximation error can be arbitrarily reduced by increasing the number of fuzzy sets.

**Example 6.2**

In this example, it is required to design a fuzzy system of the form (6.22) to approximate the function

$$g(x) = x_1 x_2 + x_2 - x_1 \qquad (6.25)$$

where $x_1 \in [-1, 1]$ and $x_2 \in [-2, 2]$. Seven and five triangular membership functions are assigned for $x_1$ and $x_2$, respectively, as shown in Figure 6.7.

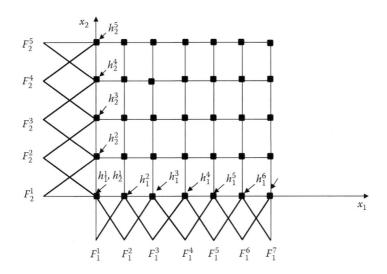

**FIGURE 6.7**
Membership functions of Example 6.2.

The parameters of these functions are $h_1^1 = -1, h_1^7 = 1,$ $\Delta_1 = 1/3,$ and $h_1^j = h_1^j + (j-1)\Delta_1$ for $j = 2, \dots, 6$. Similarly, for $x_2, h_2^1 = -2, h_2^5 = 2, \Delta_2 = 1,$ and $h_2^j = h_2^{j-1} + (j-1)\Delta_2$ for $j = 2, 3, 4$.

The fuzzy system for this example is designed as

$$f(x_1, x_2) = \frac{\sum_{i=1}^{7}\sum_{k=1}^{5} g\left(h_1^i, h_2^k\right)\mu_{F_1^i}(x_1)\mu_{F_2^k}(x_2)}{\sum_{i=1}^{7}\sum_{k=1}^{5}\mu_{F_1^i}(x_1)\mu_{F_2^k}(x_2)} \qquad (6.26)$$

where $g\left(h_1^i, h_2^k\right)$ is the function value at the designated square points in Figure 6.7. The plots of $g(x_1, x_2)$ and $f(x_1, x_2)$ are shown in Figure 6.8. It is clear that (6.26) approximates (6.25) with almost zero fuzzy approximation error.

## 6.4.2 Fuzzy Basis Function Representation of Fuzzy System

To facilitate the use of fuzzy systems in function approximation and control, the FBF expansion of (6.18) is introduced. The FBF is defined as

$$b_i(x) = \frac{\prod_{k=1}^{n} e^{-\left(\frac{x_k - \bar{x}_k^i}{\sigma_k^i}\right)^2}}{\sum_{i=1}^{L}\left(\prod_{k=1}^{n} e^{-\left(\frac{x_k - \bar{x}_k^i}{\sigma_k^i}\right)^2}\right)}, \qquad i = 1, 2, \dots, L \qquad (6.27)$$

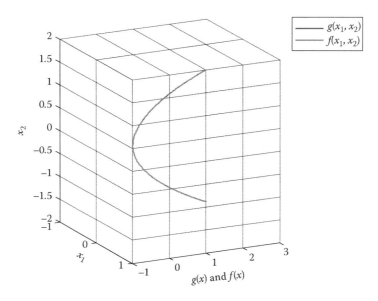

**FIGURE 6.8**
Function of Example 6.2 and its fuzzy approximation.

Then the fuzzy system (6.18) is equivalent to the following form:

$$f(x) = \sum_{i=1}^{L} b_i(x)\theta_i \qquad (6.28)$$

where $\theta_i = \bar{y}^i \in \mathfrak{R}$ are constants. The following example illustrates the advantage of using the FBFs in capturing the local and global properties of function approximation.

**Example 6.3**

Suppose for a scalar system defined in the interval $[-5,5]$, we have six fuzzy sets with Gaussian membership function $\mu_{F^i}(x) = e^{-\frac{1}{2}(x-\bar{x}^i)^2}$, where $\bar{x}^i = -5,-3,-1,1,3,5$ for $i = 1,\ldots,6$. The plot of these functions is shown in Figure 6.9. If we have six rules in the fuzzy system, that is, $L=6$, then the FBFs (6.27) will be

$$b_1(x) = \frac{e^{-\frac{1}{2}(x-\bar{x}^1)^2}}{D}, \quad b_2(x) = \frac{e^{-\frac{1}{2}(x-\bar{x}^2)^2}}{D},$$

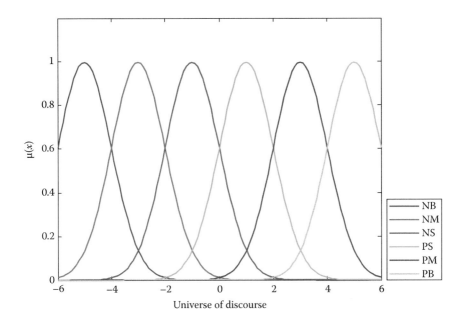

**FIGURE 6.9**
Six Gaussian membership functions for Example 6.3.

$$b_3(x) = \frac{e^{-\frac{1}{2}\left(x-\bar{x}^3\right)^2}}{D}, \quad b_4(x) = \frac{e^{-\frac{1}{2}\left(x-\bar{x}^4\right)^2}}{D},$$

$$b_5(x) = \frac{e^{-\frac{1}{2}\left(x-\bar{x}^5\right)^2}}{D}, \quad \text{and} \quad b_6(x) = \frac{e^{-\frac{1}{2}\left(x-\bar{x}^6\right)^2}}{D}$$

where $D = \sum_{i=1}^{6} e^{-\frac{1}{2}\left(x-\bar{x}^i\right)^2}$.

The plots of these FBFs are given in Figure 6.10, where the FBF property is observed. It is clear that all the FBFs with centers inside the open interval ]−5,5 [ look like Gaussian functions whereas those with centers at the boundaries of the interval look like sigmoidal functions. It is known in the neural networks literature that the Gaussian radial basis functions have the advantage of capturing the local properties of the function and the neural networks with sigmoidal nonlinearities have the advantage of capturing the global properties of the function [5]. Therefore, the FBFs combine the advantages of both types of neural networks.

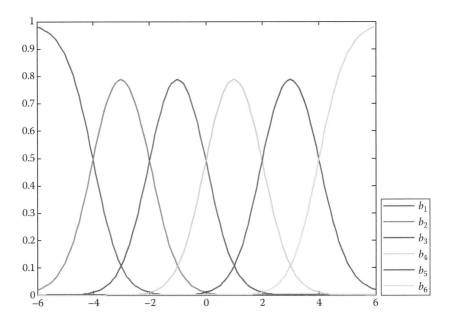

**FIGURE 6.10**
Fuzzy basis functions for Example 6.3.

# References

1. L.-X. Wang, *Adaptive Fuzzy Systems and Control: Design and Stability Analysis*, Prentice Hall, Englewood Cliffs, NJ, 1994.
2. T. Takagi and M. Sugeno, Fuzzy identification of systems and its applications on modeling and control, *IEEE Trans. Syst. Man Cybernetics*, 15(1), 116–132, 1985.
3. E. H. Mamdani, Applications of fuzzy algorithms for simple dynamic plant, *Proc. IEE*, 121(12), 175–189, 1974.
4. L.-X. Wang and J. M. Mendel, Fuzzy basis functions, universal approximation and orthogonal least-squares learning, *IEEE Trans. Neural Networks*, 3(5), 807–814, September 1992.
5. R. Lippmann, A critical overview of neural network pattern classifiers, in *Proc. 1991 IEEE Workshop on Neural Networks for Signal Processing*, Princeton, NJ, pp. 266–275, 1991.

# 7

## Nonadaptive Fuzzy Load Frequency Control

### 7.1 Introduction

Fuzzy control can be classified as nonadaptive fuzzy control and adaptive fuzzy control. In nonadaptive fuzzy control, the control parameters are fixed, while in the adaptive fuzzy control, the control parameters are time-varying and adjusted according to updating laws. In this chapter, we will present two nonadaptive fuzzy control techniques for a linear system, namely, stable and optimal fuzzy controllers. The fuzzy system used to develop these types of controllers is the one with a center average defuzzifier, product-inference rule, and a singleton fuzzifier, as described in Chapter 6.

### 7.2 Stable Fuzzy Control for SISO Linear Systems

Consider a single-input single-output (SISO) linear time-invariant system represented by the following state variable model:

$$\dot{x}(t) = Ax(t) + be(t) \tag{7.1}$$

$$y(t) = cx(t) \tag{7.2}$$

where
  $x(t) \in \mathfrak{R}^n$ is the state vector
  $e(t) \in \mathfrak{R}$ is the input
  $y(t) \in \mathfrak{R}$ is the system output

The transfer function of (7.1) and (7.2) is given by

$$\frac{Y(s)}{E(s)} = H(s) = c(sI - A)^{-1} b \tag{7.3}$$

Consider the feedback signal $v(t)$ defined by

$$v(t) = f(y(t)) \qquad (7.4)$$

where $f(y)$ is a fuzzy system with a center average defuzzifier, a product-inference rule, and a singleton fuzzifier and satisfies

$$f(0) = 0 \qquad (7.5)$$

The closed-loop system equation can be written as

$$\dot{x}(t) = Ax(t) + b(r(t) - v(t)) \qquad (7.6)$$

$$v(t) = f(y(t)) \qquad (7.7)$$

$$y(t) = cx(t) \qquad (7.8)$$

where $r(t)$ is the reference signal. A block diagram representation of the closed-loop system is shown in Figure 7.1.

The notion and condition of input–output stability is given next.

**Definition 7.1:**

Given $p : 1 \le p < \infty$, the space $L_p^n$ consists of all piecewise continuous functions $g(t) = [g_1(t), \ldots, g_n(t)]^T$: $[0, \infty) \to \Re^n$ satisfying

$$\|g(t)\|_{L_p^n} = \left( \int_0^\infty \left( |g_1|^p + |g_2|^p + \cdots + |g_n|^p \right) dt \right)^{1/p} < \infty.$$

For a special case when $p = \infty$, the space $L_\infty^n$ consists of all piecewise continuous functions $g(t)$: $[0, \infty) \to \Re^n$ satisfying $\|g(t)\|_{L_\infty^n} = \sup_{t \in [0,\infty)} \left( \max_i |g_i(t)| \right) < \infty$, $1 \le i \le n$.

A system with input $u(t) \in \Re^r$ and output $y(t) \in \Re^m$ is said to be $L_p$-stable if $u(t) \in L_p^r$ implies $y(t) \in L_p^m$. In particular, a system is $L_\infty$-stable (i.e., bounded-input bounded-output stable) if $u(t) \in L_\infty^r$ implies $y(t) \in L_\infty^m$.

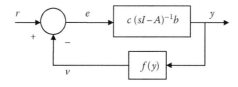

**FIGURE 7.1**
Closed-loop system with fuzzy output feedback.

**Theorem 7.1 [1]**

Suppose *f(y)* is globally Lipschitz continuous; that is, suppose there exists a finite constant $\ell$ such that

$$\left\| f\left(y_1(t)\right) - f\left(y_2(t)\right) \right\| \le \ell \left\| y_1(t) - y_2(t) \right\| \quad \text{for all} \quad y_1, y_2 \in \Re \qquad (7.9)$$

Under this condition, if the open-loop system $\dot{x}(t) = Ax(t)$ is globally exponentially stable (or equivalently, all the eigenvalues of the matrix $A$ are nonpositive), then the closed-loop system is $L_p$-stable for all $p \in [1, \infty]$.

Now, to design the fuzzy controller (7.7) that yields global exponential stability of the equilibrium point of the closed-loop, one has to design the fuzzy system *f(y(t))*. To this end, the system output $y(t) \in [a, b]$ is partitioned into $(2N+1)$ fuzzy sets where $N$ is the number of fuzzy sets to the right or to the left of the fuzzy set centered at zero. Then, $(2N+1)$ fuzzy IF-THEN rules given by

$$\text{IF } y \text{ is } F^k, \text{ THEN } u \text{ is } L^k \qquad (7.10)$$

are assumed based on the required output performance where $k = 1, 2, \dots$ , $2N+1$. The centers $\overline{y}^k$ of the fuzzy sets $L^k$ are selected according to

$$\overline{y}^k \begin{cases} \le 0 & \text{for } k = 1, \dots, N \\ = 0 & \text{for } k = N+1 \\ \ge 0 & \text{for } k = N+2, \dots, 2N+1 \end{cases} \qquad (7.11)$$

Finally, the fuzzy system is designed using a product-inference engine, a singleton fuzzifier, and a center average defuzzifier, and the feedback signal $v$ takes the form

$$v = f\left(y\right) = \frac{\sum_{k=1}^{2N+1} \overline{y}^k \mu_{F^k}\left(y\right)}{\sum_{k=1}^{2N+1} \mu_{F^k}\left(y\right)} \qquad (7.12)$$

It can be shown, by using (7.10) and (7.11), that condition (7.5) is satisfied. The fuzzy controller *f(y)* can also be shown to be a continuous, bounded, and piece-wise linear function and hence satisfies the Lipschitz condition (7.9) [2].

**Example 7.1**

Consider the load frequency control (LFC) of the isolated power area given in Example 2.1. Design a stable fuzzy logic controller for this system. The state variable model is given by

$$\dot{x} = Ax + bu + Fd \tag{7.13}$$

$$y = cx \tag{7.14}$$

where

$$A = \begin{bmatrix} -\dfrac{D}{2H} & \dfrac{1}{2H} & 0.0 \\ 0.0 & -\dfrac{1}{T_t} & \dfrac{1}{T_t} \\ -\dfrac{1}{RT_g} & 0.0 & -\dfrac{1}{T_g} \end{bmatrix}$$

$$b^T = \begin{bmatrix} 0 & 0 & \dfrac{1}{T_g} \end{bmatrix}$$

$$F^T = \begin{bmatrix} -\dfrac{1}{2H} & 0 & 0 \end{bmatrix}$$

$$c = \begin{bmatrix} 1 & 0 & 0 \end{bmatrix}$$

$d = \Delta P_L$ is the load disturbance

Using the numerical values of the parameters presented in Example 1.4, the $A$ and $b$ matrices become

$$A = \begin{bmatrix} -0.0625 & 0.0833 & 0.0 \\ 0.0 & -3.3333 & 3.3333 \\ -100 & 0.0 & -5 \end{bmatrix} \quad \text{and} \quad b = \begin{bmatrix} 0.0 \\ 0.0 \\ 5 \end{bmatrix}$$

The universe of discourse for the change of frequency (the system output) is selected as $D = [-0.01\ 0.01]$. Nine triangular membership functions are chosen, as shown in Figure 7.2 where the fuzzy sets used are NVVB (negative very very big), NVB (negative very very big), NB (negative big), Z (zero), P (positive), PB (positive big), PVB (positive very big), and PVVB (positive very very big).

The eigenvalues of the $A$ matrix are calculated as $-6.4173$, $-0.9892 \pm j1.8741$, indicating that the open-loop system is asymptotically stable, and hence the designed closed-loop system (7.6) using the fuzzy controller (7.12) will be asymptotically stable. According to (7.11), the parameters $\bar{y}^k$ are chosen as $\bar{y}^k = -10^{-3}$ for $k = 1, \dots, 4$, $\bar{y}^k = 10^{-3}$ for $k = 6, \dots, 9$, and $\bar{y}^5 = 0.0$. The closed-loop system response is shown in Figure 7.3.

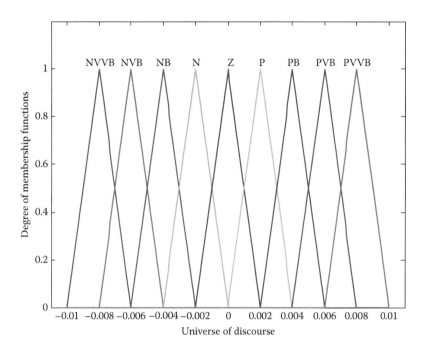

**FIGURE 7.2**
Membership functions for the LFC of an isolated area.

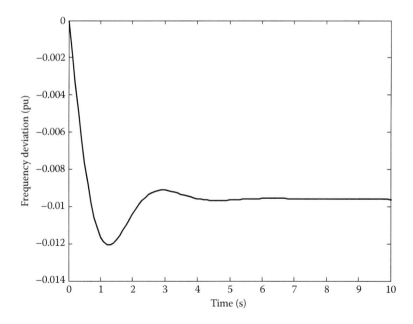

**FIGURE 7.3**
Frequency deviation for the closed-loop LFC of an isolated area with a stable fuzzy controller.

## 7.3 Optimal Fuzzy Control for SISO Linear Systems

In the previous section, we have shown how to design a stable fuzzy controller for a SISO linear system. The key point to design such a controller was to use a fuzzy system in the form (7.4) such that the closed-loop system is stable. The parameters of the fuzzy system are chosen according to (7.11) to guarantee closed-loop stability. In this section, the parameters of the fuzzy system are determined in a way to optimize a given performance index. This specific problem is termed as optimal fuzzy control. In what follows, we are going to review some basic concepts of optimal control, and then we introduce a design methodology for optimal fuzzy control.

### 7.3.1 Review of Optimal Control Concepts

In optimal control, the objective is to determine a control law $u(t)$ for the system

$$\dot{x} = g\left(x(t), u(t)\right) \tag{7.15}$$

such that the following performance index

$$J\left(x(t), t\right) = \int_{t}^{t_f} h\left(x(t), u(t)\right) d\tau \tag{7.16}$$

is minimized where $t$ is the current time, $t_f$ is the final time, $x(t)$ is the current state, and $h(x, u)$ characterizes the cost function $J(x, t)$. The control law that minimizes (7.16) and the associated trajectory are termed as the optimal control law $u^*(t)$ and the optimal trajectory $x^*(t)$, respectively.

The optimal control problem can be solved using many techniques [3–6]. One technique is to use the principle of optimality and its result known as the Hamilton–Jacobi–Bellman (HJB) equation. The principle of optimality states that "If a control is optimal from some initial state, then it must satisfy the following property: after any initial period, the control for the remaining period must also be optimal with regard to the state resulting from the control of the initial period." To apply the principle of optimality to the optimal control problem defined by (7.15) and (7.16), consider the current time $t$ and the future time $(t + \Delta t)$, and $\Delta t$ as the incremental change in time. The optimal control during the interval $[t, t + \Delta t)$ can be determined as follows. First, Equation 7.16 is rewritten as

$$J\left(x, t\right) = \int_{t}^{t+\Delta t} h\left(x, u\right) d\tau + \int_{t+\Delta t}^{t_f} h\left(x, u\right) d\tau \tag{7.17}$$

where

$$\int_{t+\Delta t}^{t_f} h(x,u)\,d\tau = J\left(x(t+\Delta t),t+\Delta t\right) \tag{7.18}$$

Therefore, the cost function can be written using (7.18) in (7.17) as

$$J(x,t) = \int_{t}^{t+\Delta t} h(x,u)\,d\tau + J\left(x(t+\Delta t),t+\Delta t\right) \tag{7.19}$$

In (7.19), $x(t+\Delta t)$ is the state at $(t+\Delta t)$ and can be approximated using Taylor's series expansion as $x(t+\Delta t) \cong x(t) + (dx(t)/dt)\Delta t \cong x(t) + g(x,t)\Delta t \cong x(t) + \Delta x$, therefore (7.19) becomes

$$J(x,t) = \int_{t}^{t+\Delta t} h(x,u)\,d\tau + J\left((x+\Delta x),t+\Delta t\right) \tag{7.20}$$

Using the principle of optimality, the optimal cost function under optimal control in the interval $[t,t+\Delta t)$ is given by

$$J^*(x,t) = \min_{u(\tau),\,t\le\tau<t+\Delta t} \left\{ \int_{t}^{t+\Delta t} h(x,u)\,d\tau + J^*\left((x+\Delta x),t+\Delta t\right) \right\} \tag{7.21}$$

where $J^*((x+\Delta x),t+\Delta t)$ is the optimal cost function at time $(t+\Delta t)$ and $J^*(x,t)$ is the optimal cost function from $(t+\Delta t)\to t$. The first and second terms on the right-hand side of (7.21) can be approximated by

$$\int_{t}^{t+\Delta t} h(x,u)\,d\tau \cong h(x,u)\,\Delta t \tag{7.22}$$

$$J^*(x+\Delta x),t+\Delta t) \cong J^*(x,t) + \left(\frac{\partial J^*}{\partial x}\right)^T \Delta x + \frac{\partial J^*}{\partial t}\Delta t \tag{7.23}$$

Substituting (7.22) and (7.23) in (7.21) and using the fact that $J^*(x,t)$ and $(\partial J^*/\partial t)\Delta t$ are independent of the control $u(\tau),t\le\tau<t+\Delta t$, we get

$$J^*(x,t) = J^*(x,t) + \frac{\partial J^*}{\partial t}\Delta t + \min_{u(\tau),\,t\le\tau<t+\Delta t} \left\{ h(x,u)\,\Delta t + \left(\frac{\partial J^*}{\partial x}\right)^T \Delta x \right\} \tag{7.24}$$

Dividing (7.24) by $\Delta t$ and replacing $\Delta x/\Delta t$ by $\dot{x}$ (since $\Delta t \to 0$), we obtain the following HJB equation

$$-\frac{\partial J^*}{\partial t} = \min_{u(t)}\left\{ h(x,u) + \left(\frac{\partial J^*}{\partial x}\right)^T g(x,u) \right\} \tag{7.25}$$

### 7.3.1.1 The Linear Quadratic Regulator Problem

Consider the linear time-invariant system given by

$$\dot{x} = Ax + Bu \tag{7.26}$$

where
$x \in \mathfrak{R}^n$
$u$ is a scalar input

The problem is to find the optimal control that minimizes the following quadratic cost function:

$$J(x,t) = \int_t^{t_f} \left( x^T Q x + u^T R u \right) d\lambda \tag{7.27}$$

where
$Q = Q^T \in \mathfrak{R}^n \geq 0$ is a symmetric and positive semi-definite matrix
$R = R^T > 0$ is a positive scalar

In this setup, $h(x(t), u(t)) = x^T Q x + u^T R u$ and $g(x(t), u(t)) = Ax + Bu$. The optimal control $u^*$ of the linear quadratic regulator problem can be found by applying the HJB equation (7.25)

$$-\frac{\partial J^*}{\partial t} = \min_{u(t)}\left\{ x^T Q x + u^T R u + \left(\frac{\partial J^*}{\partial x}\right)^T (Ax + Bu) \right\} \tag{7.28}$$

Assuming that the minimal cost is quadratic

$$J^* = x^T L(t) x \tag{7.29}$$

where $L(t) = L^T(t) \geq 0$ is a symmetric and positive, semidefinite, time-varying matrix. Then (7.28) becomes

$$-x^T \frac{\partial L(t)}{\partial t} x = \min_{u(t)}\left\{ x^T Q x + u^T R u + 2x^T L(t)(Ax + Bu) \right\} \tag{7.30}$$

The right-hand side of (7.30) is evaluated from

$$\frac{\partial}{\partial u}\left[x^T Q x + u^T R u + 2x^T L(t)(Ax + Bu)\right] = 0 \tag{7.31}$$

which gives

$$2Ru^* + 2B^T L(t)x = 0 \tag{7.32}$$

The optimal control is determined from (7.32) as

$$u^* = -R^{-1}B^T L(t)x \tag{7.33}$$

Now, using (7.33) to determine the right-hand side of (7.30) as

$$x^T Q x + u^{*T} R u^* + 2x^T L(t)(Ax + Bu^*)$$
$$= x^T\left(Q - L(t)BR^{-1}B^T L(t) + L(t)A + A^T L(t)\right)x \tag{7.34}$$

Therefore, Equation 7.30 reduces to

$$\dot{L}(t) = -Q + L(t)BR^{-1}B^T L(t) - L(t)A - A^T L(t) \tag{7.35}$$

Equation 7.35 is called the Riccati equation. When the initial time $t=0$ and the final time $t_f = \infty$, the cost function (7.27) takes the form

$$J(x) = \int_0^\infty \left(x^T Q x + u^T R u\right)d\lambda \tag{7.36}$$

In this case, the matrix $L(t)$ becomes a constant matrix and the Riccati equation (7.35) reduces to the following algebraic Riccati equation (ARE):

$$A^T L + LA + Q - LBR^{-1}B^T L = 0 \tag{7.37}$$

### 7.3.2 Optimal Fuzzy Control

In this section, we explain how a fuzzy system with center average defuzzifier, product-inference rule, and a singleton fuzzifier can be used to design an optimal fuzzy controller. Consider a linear time-invariant system of the form (7.26) where $x \in \mathfrak{R}^n$ and $u$ is a scalar input. Let $(2N_i+1)$ be the number of fuzzy sets $F_i$ representing the state $x_i$ for $i = 1, 2, \ldots, n$, where $N_i$ is the number of fuzzy sets to the right or to the left of the zero fuzzy set. The member-ship functions of each fuzzy set are denoted by $\mu_{F_i}^{\ell_i}(x_i)$. The control input $u$ is designed via the following fuzzy system:

$$u = -f(x) = -\frac{\sum_{\ell_1=1}^{2N_1+1}\cdots\sum_{\ell_n=1}^{2N_n+1}\overline{y}^{\ell_1\cdots\ell_n}\left(\prod_{i=1}^{n}\mu_{F_i}^{\ell_i}(x_i)\right)}{\sum_{\ell_1=1}^{2N_1+1}\cdots\sum_{\ell_n=1}^{2N_n+1}\left(\prod_{i=1}^{n}\mu_{F_i}^{\ell_i}(x_i)\right)} \tag{7.38}$$

The fuzzy controller (7.38) can be written in terms of fuzzy basis functions and adjustable parameters as shown in the following equation:

$$u = f_b(x)\theta(t) \tag{7.39}$$

where $f_b(x) \in \mathfrak{R}^{1\times N} = \left[f_b^1(x)f_b^2(x)\ldots f_b^N(x)\right]$ is the vector of fuzzy basis func-tions whose entries are defined by

$$f_b^k(x) = \frac{\prod_{i=1}^{n}\mu_{F_i}^{\ell_i}(x_i)}{\sum_{\ell_1=1}^{2N_1+1}\sum_{\ell_n=1}^{2N_n+1}\left(\prod_{i=1}^{n}\mu_{F_i}^{\ell_i}(x_i)\right)}, \quad k = 1,2,\ldots,N \tag{7.40}$$

where $N = \prod_{i=1}^{n}(2N_i+1)$. The vector of adjustable parameters $\theta(t)\in\mathfrak{R}^{N\times 1}$ is given by

$$\theta^T(t) = \left[-\theta^1 - \theta^2 \cdots - \theta^N\right] \tag{7.41}$$

The elements of the parameter vector $\theta^k = \overline{y}^{\ell_1\ell_2\cdots\ell_n}$, $\ell_k = 1,2,\ldots,(2N_k+1)$ should be arranged in the same order as the elements of the fuzzy basis func-tion vector. This parameter vector needs to be determined to minimize a given cost function.

As an illustration of the fuzzy basis function expansion of the controller given in (7.38), assume that $n=2$ and each state is defined by three fuzzy sets, that is, $2N_1+1=2N_2+1=3$ and $N=9$. The controller (7.38) becomes

$$u = -f(x) = -\frac{\sum_{\ell_1=1}^{3}\sum_{\ell_n=1}^{3}\overline{y}^{\ell_1\ell_2}\left(\prod_{i=1}^{2}\mu_{F_i}^{\ell_i}(x_i)\right)}{\sum_{\ell_1=1}^{3}\sum_{\ell_n=1}^{3}\left(\prod_{i=1}^{2}\mu_{F_i}^{\ell_i}(x_i)\right)} \tag{7.42}$$

which, upon expansion, gives the following:

$$u = -f(x) = \frac{1}{D} \begin{bmatrix} \mu_{F_1}^1(x_1)\mu_{F_2}^1(x_2)\,\mu_{F_1}^1(x_1)\mu_{F_2}^2(x_2) \\ \mu_{F_1}^1(x_1)\mu_{F_2}^3(x_2)\ldots\mu_{F_1}^3(x_1)\mu_{F_2}^3(x_2) \end{bmatrix} \begin{bmatrix} -\bar{y}^{11} \\ -\bar{y}^{12} \\ -\bar{y}^{13} \\ \vdots \\ -\bar{y}^{33} \end{bmatrix} \tag{7.43}$$

where $D$ is given by

$$D = \mu_{F_1}^1(x_1)\mu_{F_2}^1(x_2) + \mu_{F_1}^1(x_1)\mu_{F_2}^2(x_2) + \mu_{F_1}^1(x_1)\mu_{F_2}^3(x_2) + \mu_{F_1}^2(x_1)\mu_{F_2}^1(x_2)$$

$$+ \mu_{F_1}^2(x_1)\mu_{F_2}^2(x_2) + \mu_{F_1}^2(x_1)\mu_{F_2}^3(x_2) + \mu_{F_1}^3(x_1)\mu_{F_2}^2(x_2) + \mu_{F_1}^3(x_1)\mu_{F_2}^3(x_2).$$

Comparing (7.43) and (7.39), one can identify the fuzzy basis function vector as

$$f_b(x) = \begin{bmatrix} f_b^1(x) & f_b^2(x) \cdots f_b^N(x) \end{bmatrix}$$

where $f_b^k(x) = \dfrac{\prod_{i=1}^2 \mu_{F_i}^{\ell_i}(x_i)}{D}$, $k = 1,\ldots,9$, and the adjustable parameter vector as

$$\theta^T(t) = \begin{bmatrix} -\theta^1 - \theta^2 \cdots - \theta^N \end{bmatrix} = -\begin{bmatrix} \bar{y}^{11} & \bar{y}^{12} & \bar{y}^{13} & \bar{y}^{21} & \bar{y}^{22} & \bar{y}^{23} & \bar{y}^{31} & \bar{y}^{32} & \bar{y}^{33} \end{bmatrix}$$

To find the optimal parameter vector $\theta^*(t)$ that minimizes a quadratic cost function (7.36), we substitute the controller (7.39) in the HJB equation (7.30) to get

$$-x^T \frac{\partial L(t)}{\partial t} x = \min_{\theta(t)} \left\{ x^T Q x + (f_b\theta)^T R(f_b\theta) + 2x^T L(t)(Ax + Bf_b\theta) \right\} \tag{7.44}$$

The right-hand side of (7.44) is evaluated from

$$\frac{\partial}{\partial\theta}\left\{ x^T Q x + (f_b\theta)^T R(f_b\theta) + 2x^T L(t)(Ax + Bf_b\theta) \right\} = 0 \tag{7.45}$$

The differentiation in (7.45) yields

$$2f_b^T R f_b\theta^* = -2f_b^T B^T L(t) x^* \tag{7.46}$$

From (7.46) we get

$$f_b \theta^* = -R^{-1}B^T L(t) x^* \tag{7.47}$$

Therefore, the optimal parameter vector is determined from (7.47) as

$$\theta^* = -M_1 M_2 x^* \tag{7.48}$$

where

$$M_1 = \left( f_b f_b^T \right)^{-1} f_b^T \tag{7.49}$$

$$M_2 = R^{-1}B^T L(t) \tag{7.50}$$

Now, using the optimal parameter vector defined by (7.48) in (7.44), we get the following matrix differential equation:

$$\dot{L}(t) = -\left\{ Q + L(t)A + A^T L(t) + u^{*T} R u^* - 2L(t) B u^* \right\} \tag{7.51}$$

where

$$u^* = f_b(x) \theta^*(t) \tag{7.52}$$

Note that the last two terms on the right-hand side of (7.51) can be written as

$$u^{*T} R u^* = L(t) B R^{-1} B^T L(t) \tag{7.53}$$

and

$$2L(t) B u^* = 2L(t) B R^{-1} B^T L(t) \tag{7.54}$$

Using (7.53) and (7.54) in (7.51), we get

$$\dot{L}(t) = -Q + L B R^{-1} B^T L - L(t) A - A^T L(t) \tag{7.55}$$

For infinite final time, we obtain the following ARE:

$$Q - L B R^{-1} B^T L + L A + A^T L = 0 \tag{7.56}$$

The procedure to design an optimal fuzzy controller of the form (7.52) consists of two stages, namely, an off-line and an online stages. These stages are summarized in the following.

*Off-line computation:*

1. For the state variable $x_i$, select $(2N_i+1)$ fuzzy sets $F_i$ to cover the entire universe of discourse, where $i=1,2,\ldots,n$.

2. Determine the elements of the fuzzy basis function vector $f_b^k(x)$ using (7.40).

3. Solve the ARE (7.55) to find the constant matrix $L$.

4. The optimal parameter vector is determined using (7.48).

*Online computation:*

The optimal fuzzy control signal is determined using (7.56).

The following example illustrates the foregoing procedure to design an optimal fuzzy load frequency controller for a single-area system.

**Example 7.2**

Consider the data in the previous example. Assume that the state vector is perturbed from the steady state such that the initial state vector is given by $x^T=[-0.05 \quad 0 \quad 0]$. This means that the initial frequency deviation is −5%. Each state variable is represented by five fuzzy sets, that is, $N_1=N_2=N_3=2$. The membership functions are selected as

$\mu_{F_i}^{\ell_i}(x_i)=e^{-\left(\left(x_i-\bar{x}_i^{\ell_i}\right)/\sigma_i^{\ell_i}\right)^2}$. Centers and widths $(\bar{x}_{1,}^{\ell_1} \quad \sigma_1^{\ell_1})$ of the membership functions for $x_1$ are shown in Table 7.1 and those for $x_2$ and $x_3$ are assumed equal and shown in Table 7.2.

The $Q$ matrix is assumed identity and the scalar $R=1.0$. There are $N=(2N_1+1)(2N_2+1)(2N_3+1)=125$ fuzzy basis functions; some of these functions are shown in Figure 7.4.

**TABLE 7.1**

Parameters of the Membership Functions for $x_1$

| $\ell_1$ | $\bar{x}_1^{\ell_1}$ | $\sigma_1^{\ell_1}$ |
|---|---|---|
| 1 | −0.01 | 0.0031 |
| 2 | −0.005 | 0.0031 |
| 3 | 0 | 0.0031 |
| 4 | 0.005 | 0.0031 |
| 5 | 0.01 | 0.0031 |

**TABLE 7.2**

Parameters of the Membership Functions for $x_2$ and $x_3$

| $\ell_2=\ell_3$ | $\overline{x}_2^{\ell_2}=\overline{x}_3^{\ell_3}$ | $\sigma_2^{\ell_2}=\sigma_3^{\ell_3}$ |
|---|---|---|
| 1 | −0.4 | 0.1 |
| 2 | −0.2 | 0.1 |
| 3 | 0 | 0.1 |
| 4 | 0.2 | 0.1 |
| 5 | 0.4 | 0.1 |

Solution of the ARE gives $L = \begin{bmatrix} 196.97 & 3.04 & -0.12 \\ 3.04 & 0.2115 & 0.06 \\ -0.12 & 0.06 & 0.11 \end{bmatrix}$. The vector of

optimal parameters is calculated from (7.48) and it has 125 entries; some of them are shown in Figure 7.5.

The optimal fuzzy controller defined by (7.52) is used to simulate the closed-loop response of the isolated LFC area subject to the initial conditions $x^T = [-0.05 \quad 0 \quad 0]$. The simulation results are shown in Figure 7.6 and the optimal fuzzy control signal is shown in Figure 7.7.

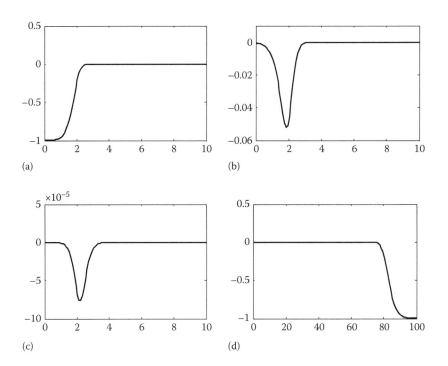

**FIGURE 7.4**

Fuzzy basis functions: (a) $f_b^1(x)$, (b) $f_b^{26}(x)$, (c) $f_b^{51}(x)$, and (d) $f_b^{125}(x)$.

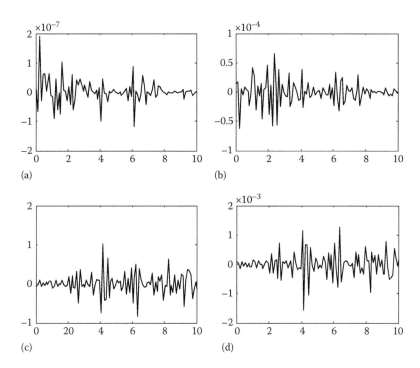

**FIGURE 7.5**
Optimal parameters: (a) $\theta^{*1}$, (b) $\theta^{*26}$, (c) $\theta^{*75}$, and (d) $\theta^{*99}$.

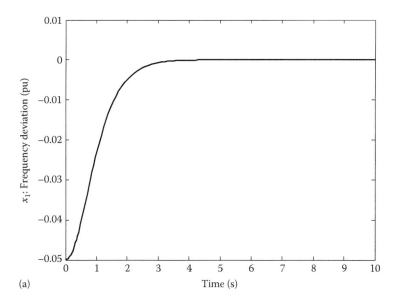

**FIGURE 7.6**
Simulation results under optimal fuzzy controller: (a) $x_1$. *(Continued)*

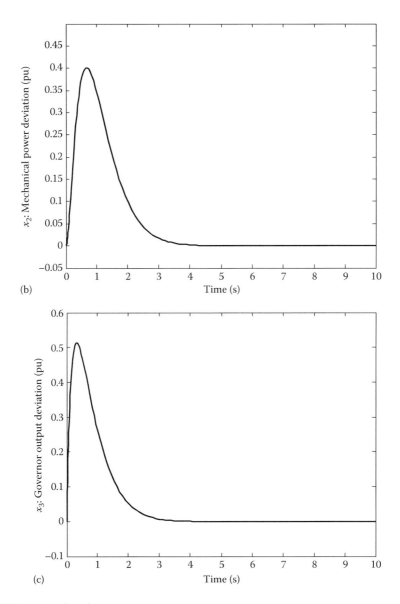

(b)

(c)

**FIGURE 7.6 (*Continued*)**
Simulation results under optimal fuzzy controller: (b) $x_2$, and (c) $x_3$.

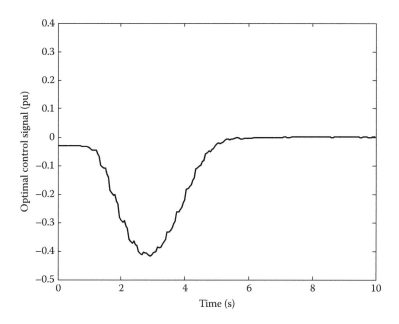

**FIGURE 7.7**
Optimal fuzzy control signal.

# References

1. M. Vidyasagar, *Nonlinear Systems Analysis*, 2nd edn., Prentice Hall, Englewood Cliffs, NJ, 1993.
2. L.-X. Wang, *A Course in Fuzzy Systems and Control*, Prentice-Hall International, Inc., Englewood Cliffs, NJ, 1997.
3. R. S. Burns, *Advanced Control Engineering*, Butterworth-Heinemann, Oxford, MA, 2001.
4. D. E. Kirk, *Optimal Control Theory: An Introduction*, Dover Publication Inc., New York, 1998.
5. D. Liberzon, *Calculus of Variations and Optimal Control Theory: A Concise Introduction*, Princeton University Press, Englewood Cliffs, NJ, 2012.
6. F. Lin, *Robust Control Design: An Optimal Control Approach*, John Wiley & Sons Ltd, West Sussex, U.K., 2007.

# 8

## Adaptive Fuzzy Control Techniques

## 8.1 Introduction

The fuzzy logic systems described in Chapter 6 can be used as controllers. When there are uncertainties or unknown variations in plant parameters, the need to use an adaptive fuzzy logic controllers is obvious. An adaptive fuzzy logic controller consists of a fuzzy logic system plus a parameter adaptation mechanism. Linguistic fuzzy information available from a human operator can be incorporated easily into the adaptive fuzzy controller. This is considered as the main advantage of adaptive fuzzy control over conventional adaptive control. The types of adaptive fuzzy controllers and their use in tracking control are explained in this chapter.

## 8.2 Types of Adaptive Fuzzy Controllers

Adaptive fuzzy controllers can be classified as direct and indirect. In direct adaptive fuzzy control, the fuzzy logic system can be used as controllers. In this type, linguistic fuzzy control rules can be directly incorporated into the controller. Indirect adaptive fuzzy controllers employ the fuzzy logic system to model an unknown plant and then the controllers are constructed based on the certainty equivalent principle. In this type, fuzzy IF-THEN rules describing the plant can be incorporated.

Another classification of adaptive fuzzy controller depends on whether the fuzzy logic system is linear or nonlinear in its adjustable parameters. In this regard, adaptive fuzzy controllers are classified into two categories. When the adjustable parameters appear linearly in the fuzzy system, the adaptive fuzzy controller is termed as first-type adaptive fuzzy controller. The following fuzzy system is used in the first type of adaptive fuzzy controller:

$$f(x) = \frac{\sum_{i=1}^{L} \bar{y}^i \left( \prod_{k=1}^{n} \mu_{F_k^i}(x_k) \right)}{\sum_{i=1}^{L} \left( \prod_{k=1}^{n} \mu_{F_k^i}(x_k) \right)} = \theta^T \phi(x) \tag{8.1}$$

where
$\theta^T(t) = \begin{bmatrix} \bar{y}^1 \bar{y}^2 \ldots \bar{y}^L \end{bmatrix}$ and $\phi(x) = [\varphi_1(x) \ \varphi_2(x) \ldots \varphi_L(x)]^T$ are the vector of adjustable parameters and the vector of fuzzy basis functions, respectively.
Each fuzzy basis function is defined by

$$\varphi_i(x) = \frac{\prod_{k=1}^{n} \mu_{F_k^i}(x_k)}{\sum_{i=1}^{L} \left( \prod_{k=1}^{n} \mu_{F_k^i}(x_k) \right)} \tag{8.2}$$

where $\mu_{F_k^i}(x_k)$ are the membership functions. The membership functions are assumed unchanged during the adaptation process.

On the other hand, if the adjustable parameters appear nonlinearly in the fuzzy logic system, the adaptive fuzzy logic system is termed as second-type adaptive fuzzy control. The fuzzy system given by (8.3) is used in the second type of adaptive fuzzy logic controller

$$f(x) = \frac{\sum_{i=1}^{L} \bar{y}^i \left( \prod_{k=1}^{n} e^{-\left( \frac{x_k - \bar{x}_k^i}{\sigma_k^i} \right)^2} \right)}{\sum_{i=1}^{L} \left( \prod_{k=1}^{n} e^{-\left( \frac{x_k - \bar{x}_k^i}{\sigma_k^i} \right)^2} \right)} \tag{8.3}$$

where $\bar{y}^i, \bar{x}_k^i$, and $\sigma_k^i$ are adjustable parameters. In this book, we consider only the first type of adaptive fuzzy controller.

In the next section, we present systematic procedures to employ fuzzy logic systems to design direct and indirect adaptive fuzzy tracking controllers. First, we review the basic concepts of a tracking controller design.

## 8.3 Tracking Controller

Consider an $n$th order, single-input single-output linear system described by the following state model:

$$\left. \begin{array}{l} \dot{x} = Ax + Bu \\ y = Cx \end{array} \right\} \tag{8.4}$$

where

$x = [x_1 \dots x_n]^T$, $u$ and $y$ represent the system states, input and output, respectively; the matrices $A \in \Re^{n \times n}$, $B \in \Re^{n \times 1}$, and $C \in \Re^{1 \times n}$ are in the following controller canonical form:

$$A = \begin{bmatrix} 0 & 1 & 0 & \cdots & 0 \\ 0 & 0 & 1 & \cdots & 0 \\ \vdots & \vdots & \vdots & \ddots & \vdots \\ 0 & 0 & 0 & 0 & 1 \\ -a_0 & -a_1 & -a_2 & \cdots & -a_{n-1} \end{bmatrix}, \quad B = \begin{bmatrix} 0 \\ 0 \\ \vdots \\ \vdots \\ 1 \end{bmatrix}, \quad \text{and} \quad C = \begin{bmatrix} c_0 & c_1 & \cdots & c_m & 0 \cdots 0 \end{bmatrix}$$

with constant parameters $a_i$, $i = 0, \dots, n-1$, $c_i$, $i = 0, \dots, m$, and $c_m \neq 0$.

The approach to design a tracking controller is based on the notion of the relative degree that can be shown as follows. From (8.4), the first derivative of the output can be written as $\dot{y} = CAx + CBu$. Since $CB = 0$, then $\dot{y} = CAx$. This means that the input $u$ does not appear in $\dot{y}$ and therefore the relative degree is not one. The second derivative is determined as $\ddot{y} = CA^2x + CABu = CA^2x$, where $CAB = 0$. Upon repetitive differentiation until the input appears, we get

$$y^{(n-m)} = CA^{(n-m)}x + CA^{(n-m-1)}Bu \tag{8.5}$$

where $y^{(n-m)}$ is the $(n-m)$th derivative of the output. Using the matrices $A$, $B$, and $C$ defined earlier, we observe that

$$CA = \begin{bmatrix} \underbrace{c_0 \quad c_1 \quad \cdots \quad c_m}_{m+1} & \underbrace{0 \cdots 0}_{n-m-1} \end{bmatrix} \begin{bmatrix} 0 & 1 & 0 & \cdots & 0 \\ 0 & 0 & 1 & \cdots & 0 \\ \vdots & \vdots & \vdots & \ddots & \vdots \\ 0 & 0 & 0 & 0 & 1 \\ -a_0 & -a_1 & -a_2 & \cdots & -a_{n-1} \end{bmatrix}$$

$$= \begin{bmatrix} \underset{1}{0} & c_0 & c_1 & \cdots & c_m & 0 \cdots 0 \end{bmatrix},$$

$$CA^2 = \begin{bmatrix} \underbrace{0 \quad 0}_{2} & c_0 & c_1 & \cdots & c_m & 0 \cdots 0 \end{bmatrix}$$

and so on to get

$$CA^{(n-m-1)} = \begin{bmatrix} \underbrace{0 \cdots 0}_{n-m-1} & c_0 & c_1 & \cdots & c_m \end{bmatrix}.$$

Hence, we obtain the following matrices:

$$CA^{(n-m)} = \begin{bmatrix} -a_0 c_m - a_1 c_m - a_2 c_m \cdots - a_{n-1} c_m \end{bmatrix} \tag{8.6}$$

$$CA^{(n-m-1)}B = c_m \neq 0 \tag{8.7}$$

Since the input appears in the $(n-m)$th derivative of the output, the relative degree of the system is $(n-m)$. Note that the transfer function realization of the system (8.4) is obtained as

$$G(s) = \frac{c_m s^m + c_{m-1} s^{m-1} + \cdots c_1 s + c_0}{s^n + a_{n-1} s^{n-1} + a_{n-2} s^{n-2} + \cdots a_1 s + a_0} \tag{8.8}$$

This means that the relative degree is indeed the same as the difference between the number of plant poles $n$ and the number of plant zeros $m$.

The objective of the tracking controller is to force the output $y$ to track a given reference signal $y_r$ as $t \to \infty$. The $(n-m)$th derivative of the reference signal is assumed to be bounded. The tracking problem can be transformed to a stabilization problem [1,2] by defining the tracking error $e = y - y_r$ and introducing the following variables:

$$
\left.\begin{aligned}
e_1 &= e \\
e_2 &= \dot{e}_1 \\
&\vdots \\
e_{n-m} &= e_1^{(n-m-1)}
\end{aligned}\right\} \tag{8.9}
$$

Taking the first derivative of both sides of (8.9) gives

$$
\left.\begin{aligned}
\dot{e}_1 &= e_2 \\
\dot{e}_2 &= e_3 \\
&\vdots \\
\dot{e}_{n-m} &= e_1^{(n-m)} = y_r^{(n-m)} - y^{(n-m)}
\end{aligned}\right\} \tag{8.10}
$$

The set of the first-order differential equations (8.10) can be written in the following compact form:

$$
\dot{\bar{e}} = \Lambda \bar{e} + B_c \left[ \left( y_r^{(n-m)} - CA^{(n-m)}x \right) - c_m u \right] \tag{8.11}
$$

where

$$
\bar{e} = \begin{bmatrix} e_1 & \cdots & e_{n-m} \end{bmatrix}^T
$$

$$
\Lambda = \begin{bmatrix}
0 & 1 & 0 & \cdots & 0 \\
0 & 0 & 1 & \cdots & 0 \\
\vdots & \vdots & \ddots & \ddots & \vdots \\
0 & 0 & 0 & 0 & 1 \\
0 & 0 & 0 & \cdots & 0
\end{bmatrix} \text{ is an } (n-m) \times (n-m) \text{ matrix}
$$

$$
B_c = \begin{bmatrix} 0 \\ 0 \\ \vdots \\ \vdots \\ 1 \end{bmatrix} \text{ is an } (n-m) \times 1 \text{ matrix}
$$

The control signal

$$
u = \frac{1}{c_m} \left( y_r^{(n-m)} - CA^{n-m}x + v \right) \tag{8.12}
$$

converts the system (8.11) to the linear time-invariant system in Brunovsky form:

$$\dot{\bar{e}} = \Lambda \bar{e} + B_c\left(-v\right) \tag{8.13}$$

where
  the pair $(\Lambda, B_c)$ is controllable
  $v$ is an external input signal

Note that the feedback loop consists of two components $u$ and $v$. The purpose of the control signal $u$ is to put the system (8.11) in Brunovsky form while the objective of the control signal $v$ is to stabilize the resulting linear system (8.13). Therefore, the control (8.12) transforms the tracking problem to the stabilization of the system (8.13). The stabilizing controller can be designed as a state feedback controller of the form

$$v = \begin{bmatrix} k_0 & k_1 & \cdots & k_{n-m-1} \end{bmatrix} \begin{bmatrix} e_1 \\ e_2 \\ \vdots \\ e_{n-m} \end{bmatrix} = K^T \bar{e} \tag{8.14}$$

Substituting (8.14) in (8.13) leads to the following closed-loop system:

$$\dot{\bar{e}} = \Lambda_c \bar{e} \tag{8.15}$$

where $\Lambda_c = (\Lambda - B_c K^T)$. The characteristic polynomial of (8.15) is defined by

$$\left| sI - \Lambda_c \right| = s^{n-m} + k_{n-m-1}s^{n-m-1} + \cdots + k_2 s^2 + k_1 s + k_0 \tag{8.16}$$

Equation 8.16 can be shown in another way as follows. Since the matrix

$$\Lambda_c = \left(\Lambda - B_c K^T\right) = \begin{bmatrix} 0 & 1 & 0 & \cdots & 0 \\ 0 & 0 & 1 & \cdots & 0 \\ \vdots & \vdots & \vdots & \vdots & \vdots \\ 0 & 0 & 0 & 0 & 1 \\ -k_0 & -k_1 & -k_2 & \cdots & -k_{n-m-1} \end{bmatrix}$$

then the last equation of (8.15) becomes

$$\dot{e}_{n-m} = -k_0 e_1 - k_1 e_2 - k_2 e_3 \cdots - k_{n-m-1} e_{n-m} \tag{8.17}$$

Using the definitions shown in (8.9) yields

$$e^{(n-m)} = -k_0 e - k_1 \dot{e} - k_2 \ddot{e} \cdots - k_{n-m-1} e^{(n-m-1)} \tag{8.18}$$

which leads to

$$\left( s^{n-m} + k_{n-m-1} s^{n-m-1} + \cdots + k_2 s^2 + k_1 s + k_0 \right) E(s) = 0 \tag{8.19}$$

where $E(s)$ is the Laplace transform of the error signal $e$. Therefore, the characteristic polynomial of (8.19) is the same as (8.16). The feedback gain matrix $K$ is selected such that the characteristic polynomial (8.16) is Hurwitz.

## 8.4 Direct Adaptive Fuzzy Tracking Controller

Suppose the plant to be controlled is represented by

$$\left. \begin{array}{l} \dot{x} = Ax + Bu \\ y = Cx \end{array} \right\} \tag{8.20}$$

where the matrices $(A, B, C)$ are in controller canonical forms defined earlier, the entries of the last row of matrix $A$ are unknown and $c_m$ is a known non-zero constant.

The control objective is to design a control law

$$u = u_d(x, \theta) \tag{8.21}$$

using fuzzy logic system such that the plant output $y(t)$ tracks a reference trajectory $y_r(t)$. The adjustable parameter vector $\theta$ is updated such that the tracking error reaches zero asymptotically.

### 8.4.1 Design of Fuzzy Controller

Some human control experience may be available in the form of IF-THEN rules

$$\text{IF} \quad x_1 \text{ is } H_1^k \text{ and } x_2 \text{ is } H_2^k \cdots \text{ and } x_n \text{ is } H_n^k \quad \text{THEN } u \text{ is } C^k \tag{8.22}$$

where $H_i^k$ and $C^k$ are fuzzy sets, $k=1,\ldots,L_u$, and $L_u$ is the number of fuzzy rules. To incorporate such rules, a fuzzy system with product-inference engine, singleton fuzzifier, and center average defuzzifier of the form given by (7.38) is used as a controller. This means that, the direct adaptive fuzzy controller is given by

$$u_d\left(x,\theta\right) = \frac{\sum_{\ell_1=1}^{m_1}\cdots\sum_{\ell_n=1}^{m_n}\bar{y}^{\ell_1\cdots\ell_n}\left(\prod_{i=1}^{n}\mu_{F_i}^{\ell_i}\left(x_i\right)\right)}{\sum_{\ell_1=1}^{m_1}\cdots\sum_{\ell_n=1}^{m_n}\left(\prod_{i=1}^{n}\mu_{F_i}^{\ell_i}\left(x_i\right)\right)} \tag{8.23}$$

The following two steps provide a procedure to construct the fuzzy system $u_d(x,\theta)$:

1. For the state variable $x_i$, choose $m_i$ fuzzy sets $F_i^{\ell_i}$, $\ell_i=1,2,\ldots,m_i$ to cover the entire universe of discourse where $i=1,2,\ldots,n$.
2. Construct the fuzzy system (8.23) from the rules.

IF   $x_1$ is $F_1^{\ell_1}$ and $x_2$ is $F_2^{\ell_2}\cdots$ and $x_n$ is $F_n^{\ell_n}$   THEN $u$ is $P^{\ell_1\ell_2\cdots\ell_n}$   (8.24)

where $P^{\ell_1\ell_2\cdots\ell_n}$ is the fuzzy set of the control signal $u_d$. The number of rules in the form (8.24) is $N=m_1\times m_2\times\cdots\times m_n$. The way to find the fuzzy sets $P^{\ell_1\ell_2\cdots\ell_n}$ is as follows:

1. When the IF part of (8.22) agrees with the IF part of (8.24), then $P^{\ell_1\ell_2\cdots\ell_n}=C^k$.
2. Otherwise, $P^{\ell_1\ell_2\cdots\ell_n}$ is chosen as an arbitrary fuzzy set.

Note that the fuzzy system (8.23) can be decomposed into a vector of fuzzy basis functions multiplied by a vector of adjustable parameters, as has been shown previously in Chapter 7 (Equation 7.39). Therefore, the direct fuzzy controller $u_d(x,\theta)$ of (8.23) can be rewritten as

$$u_d\left(x,\theta\right) = \theta^T\left(t\right)f_b\left(x\right) \tag{8.25}$$

where

$f_b\left(x\right)\in\Re^{N\times 1} = \left[f_b^1\left(x\right)\ f_b^2\left(x\right)\cdots f_b^N\left(x\right)\right]^T$ and $\theta(t)\in\Re^{N\times 1}$ are the vectors of fuzzy basis functions and adjustable parameters, respectively.

Each entry $f_b^k(x)$ is defined by

$$f_b^k(x) = \frac{\prod_{i=1}^{n} \mu_{F_i}^{\ell_i}(x_i)}{\sum_{\ell_1=1}^{m_1} \cdots \sum_{\ell_n=1}^{m_n} \left( \prod_{i=1}^{n} \mu_{F_i}^{\ell_i}(x_i) \right)}, \quad k = 1, 2 \ldots, N \quad (8.26)$$

and the entries of the adjustable parameter vector are given as

$$\theta^T(t) = \begin{bmatrix} \theta^1 \theta^2 \cdots \theta^N \end{bmatrix} \quad (8.27)$$

The updating laws for these parameters will be derived in the next section.

### 8.4.2 Design of Parameter Adaptation Law

In this section, Lyapunov synthesis approach is employed to develop the adaptive law to update the parameters of the fuzzy systems. First, we determine the error dynamic equation of the system and then choose a positive definite Lyapunov function. To ensure stability of the error equation, the time derivative of the Lyapunov function has to be negative definite. This condition will impose the updating rules of the parameters. Assume that the plant parameters are fully known. In this case, the control law of (8.12) is termed as the ideal control. This control provides a linear time-invariant system (8.13), and hence a tracking controller can be easily designed. However, if the parameters are unknown, the direct adaptive fuzzy controller of the form (8.23) can be used to achieve asymptotic tracking performance, provided the parameters of the fuzzy system are updated according to specific adaptation laws. To this effect, substitute the fuzzy controller $u_d(x, \theta)$ into (8.11) to get

$$\dot{\bar{e}} = \Lambda \bar{e} + B_c \left[ \left( y_r^{(n-m)} - CA^{(n-m)}x \right) - c_m u_d(x, \theta) \right] \quad (8.28)$$

The control law of (8.12) can be written, by substituting (8.14) into (8.12), as

$$u^* = \frac{1}{c_m} \left( y_r^{(n-m)} - CA^{n-m}x + K^T \bar{e} \right) \quad (8.29)$$

By adding and subtracting the term $c_m u^*$ to the right-hand side of (8.28), we obtain

$$\dot{\bar{e}} = \Lambda_c \bar{e} + B_c c_m \left[ u^* - u_d(x, \theta) \right] \quad (8.30)$$

Equation 8.30 represents the tracking error dynamics when the fuzzy controller is used. The term $[u^* - u_d(x,\theta)]$ represents the mismatch between the ideal control and the direct fuzzy control. This mismatch has a minimum value

$$w_u = \left[ u^* - u_d\left(x,\theta^*\right) \right] \qquad (8.31)$$

where $\theta^*$ is the optimal parameter defined by

$$\theta^* = \arg\min_{\theta \in \Re^N} \left[ \sup_{x \in \Re^n} \left| u^* - u_d\left(x,\theta\right) \right| \right] \qquad (8.32)$$

where $\left( \arg\min_{\theta}\left(h(\theta)\right) \right)$ stands for argument of the minimum, which returns the value of $\theta$ such that $h(\theta)$ attains its minimum. The term $w_u$ is termed as the minimum fuzzy approximation error. Adding and subtracting the term $u_d(x,\theta^*)$ to the right-hand side of (8.30) leads to

$$\dot{e} = \Lambda_c \bar{e} + B_c c_m \left[ u_d\left(x,\theta^*\right) - u_d\left(x,\theta\right) \right] + B_c c_m w_u \qquad (8.33)$$

Using the expansion of $u_d(x,\theta)$ given in (8.25), the preceding equation (8.33) becomes

$$\dot{e} = \Lambda_c \bar{e} + c_m \left[ \psi^T f_b + w_u \right] B_c \qquad (8.34)$$

where $\psi^T = (\theta^* - \theta)^T$ is the parameter error vector.

Now, to investigate the behavior of the tracking error dynamic equation (8.34), we choose the following positive definite Lyapunov function:

$$V = \frac{1}{2}\bar{e}^T P \bar{e} + \frac{1}{2\gamma}\psi^T \psi \qquad (8.35)$$

where
$\gamma$ is a positive constant
$P$ is a positive definite symmetric matrix satisfying the Lyapunov equation

$$\Lambda_c^T P + P \Lambda_c^T = -Q \qquad (8.36)$$

with arbitrary positive definite matrix $Q$. The time derivative of (8.35) is determined as

$$\dot{V} = \frac{1}{2}\left( \dot{\bar{e}}^T \, P\bar{e} + \bar{e}^T P \dot{\bar{e}} \right) + \frac{1}{\gamma}\psi^T \dot{\psi} \qquad (8.37)$$

Using (8.34), the first two terms on the right-hand side of (8.37) become

$$\frac{1}{2}\left(\dot{\bar{e}}^T P\bar{e} + \bar{e}^T P\dot{\bar{e}}\right) = \frac{1}{2}\left(\bar{e}^T \Lambda_c^T + c_m\left(\psi^T f_b + w_u\right)B_c^T\right)P\bar{e}$$

$$+\frac{1}{2}\bar{e}^T P\left(\Lambda_c \bar{e} + c_m\left(\psi^T f_b + w_u\right)B_c\right) \quad (8.38)$$

which reduces to

$$\frac{1}{2}\left(\dot{\bar{e}}_T P\bar{e} + \bar{e}^T P\dot{\bar{e}}\right) = \frac{1}{2}\bar{e}^T\left(\Lambda_c^T P + P\Lambda_c\right)\bar{e} + c_m\left(\psi^T f_b + w_u\right)\bar{e}^T P B_c \quad (8.39)$$

The third term on the right hand side of (8.37) is determined as

$$\frac{1}{\gamma}\psi^T\dot{\psi} = -\frac{1}{\gamma}\psi^T\dot{\theta} \quad (8.40)$$

The facts that $c_m(\psi^T f_b + w_u)$ and $\bar{e}^T P B_c$ are scalars and $\theta^*$ is constant are used to reach to (8.39) and (8.40). Therefore, the time derivative of the Lyapunov function becomes

$$\dot{V} = \frac{1}{2}\bar{e}^T\left(\Lambda_c^T P + P\Lambda_c\right)\bar{e} + \psi^T\left(c_m\bar{e}^T P B_c f_b - \frac{1}{\gamma}\dot{\theta}\right) + c_m w_u\bar{e}^T P B_c \quad (8.41)$$

The parameter updating law is chosen to nullify the middle term of (8.41), that is,

$$\dot{\theta} = \gamma c_m\bar{e}^T P B_c f_b \quad (8.42)$$

Using (8.36) and (8.42), we write (8.41) as

$$\dot{V} = -\frac{1}{2}\bar{e}^T Q\bar{e} + c_m w_u\bar{e}^T P B_c \quad (8.43)$$

Taking the norm of (8.43) leads to

$$\dot{V} \le -\frac{1}{2}\lambda_{\min}(Q)\|\bar{e}\|^2 + c_m\|w_u\|\|P B_c\|\|\bar{e}\| \quad (8.44)$$

$$\dot{V} \leq -\frac{1}{2}\lambda_{\min}(Q)\|\bar{e}\|^2 + \alpha\lambda_{\max}(P)\|\bar{e}\| \qquad (8.45)$$

where

$\|B_c\| = 1$

$\alpha$ is a positive constant such that $c_m\,\|w_u\| \leq \alpha$

$\|.\|$ stands for the Euclidean norm

$\lambda_{\min}(.)$ and $\lambda_{\max}(.)$ are the minimum and maximum eigenvalues of the indicated matrix

The preceding inequality can be put in the following form:

$$\dot{V} \leq -\frac{1}{2}\lambda_{\min}(Q)(1-\beta)\|\bar{e}\|^2 - \left[\lambda_{\min}(Q)\beta\|\bar{e}\| - \alpha\lambda_{\max}(P)\right]\|\bar{e}\| \qquad (8.46)$$

where $0<\beta<1$. Provided that

$$\|\bar{e}\| \geq \frac{\alpha\lambda_{\max}(P)}{\beta\lambda_{\min}(Q)} = r \qquad (8.47)$$

Equation 8.46 becomes

$$\dot{V} \leq -\lambda_{\min}(Q)(1-\beta)\|\bar{e}\|^2 \qquad (8.48)$$

In this equation, $\dot{V} \leq 0$ and therefore the tracking error $\bar{e}$ is globally ultimately bounded. To find the ultimate bound of the tracking error, we use theorem (4.18) presented in [3]. The Lyapunov function given in (8.35) can be written as the following two inequalities:

$$\frac{1}{2}\lambda_{\min}(P)\|\bar{e}\|^2 \leq V \leq \frac{1}{2}\lambda_{\max}(P)\|\bar{e}\|^2 \qquad (8.49)$$

The ultimate bound can be determined as

$$b = \mu_1^{-1} \circ \mu_2(r) \qquad (8.50)$$

where

$\mu_1 = \dfrac{1}{2}\lambda_{\min}(P)\|\bar{e}\|^2$

$\mu_2 = \dfrac{1}{2}\lambda_{\max}(P)\|\bar{e}\|^2$

$\circ$ stands for function composition

The function inverse $\mu_1^{-1}$ is computed as $\mu_1^{-1} = \sqrt{\dfrac{2\|\bar{e}\|}{\lambda_{min}(P)}}$, and the ultimate bound using function composition is given by

$$b = \mu_1^{-1} \circ \mu_2(r) = \sqrt{\frac{2\mu_2(r)}{\lambda_{min}(P)}} = \sqrt{\frac{\lambda_{max}(P)r^2}{\lambda_{min}(P)}} = \frac{\alpha\lambda_{max}(P)}{\beta\lambda_{min}(Q)}\sqrt{\frac{\lambda_{max}(P)}{\lambda_{min}(P)}} \quad (8.51)$$

Therefore, the tracking error is ultimately bounded by the value $b$ given in (8.51), which is a function of maximum, minimum eigenvalues of $P$, minimum eigenvalues of $Q$, and the upper bound of the mismatch $w_u$.

### 8.4.3 Attenuation of Tracking Error

Tracking error can be attenuated to a desired level by introducing an attenuation control part $u_a$ to the direct adaptive fuzzy control law (8.25). Therefore, the new control law that achieves a desired tracking error is given by

$$u = u_d(x,\theta) + u_a \quad (8.52)$$

The next analysis shows how to select the control part $u_a$. To this end, replace $u_d(x,\theta)$ in (8.30) by (8.52) to get

$$\dot{\bar{e}} = \Lambda_c\bar{e} + B_c c_m \left[ u^* - u_d(x,\theta) - u_a \right] \quad (8.53)$$

Then, using the same procedure of the previous section, we reach

$$\dot{\bar{e}} = \Lambda_c\bar{e} + c_m\left(\psi^T f_b + w_u - u_a\right)B_c \quad (8.54)$$

Now, we state the following theorem.

### Theorem 8.1

If $c_m$ is an assumed known nonzero constant and the attenuation control part is chosen according to

$$u_a = \frac{1}{c_m\sigma}\bar{e}^T PB_c \quad (8.55)$$

where $\sigma$ is a positive constant, then the direct adaptive fuzzy control given by (8.25), (8.52), and (8.55) along with the adaptation law (8.42) ensures the boundedness of the tracking error and the parameter error ψ in the sense of the following $H_\infty$ tracking performance index [4]

$$\int_0^t \bar{e}^T Q \bar{e} d\tau \leq 2V(0) + \rho^2 \int_0^t w_u^2 d\tau \tag{8.56}$$

where
  ρ is the attenuation level
  $P$ is a positive definite symmetric matrix satisfying the Riccati-like equation

$$\Lambda_c^T P + P\Lambda_c - PB_c \left( \frac{2}{\sigma} - \frac{1}{\rho^2} \right) B_c^T P = -Q \tag{8.57}$$

*Proof:*

Consider the Lyapunov function (8.35). Then, its time derivative along the trajectories (8.42) and (8.54) becomes

$$\dot{V} = \frac{1}{2} \bar{e}^T \left( \Lambda_c^T P + P\Lambda_c \right) \bar{e} + w_u \bar{e}^T PB_c - c_m u_a \bar{e}^T PB_c \tag{8.58}$$

Using (8.55) in (8.58), we get

$$\dot{V} = \frac{1}{2} \bar{e}^T \left( \Lambda_c^T P + P\Lambda_c \right) \bar{e} + w_u \bar{e}^T PB_c - \frac{1}{\sigma} \bar{e}^T PB_c B_c^T P\bar{e} \tag{8.59}$$

Adding and subtracting the term $\left( 1/2\rho^2 \right) \bar{e}^T PB_c B_c^T P\bar{e}$ to the right-hand side of (8.59) and rearranging, we obtain

$$\dot{V} = \frac{1}{2} \bar{e}^T \left[ \Lambda_c^T P + P\Lambda_c - PB_c \left( \frac{2}{\sigma} - \frac{1}{\rho^2} \right) B_c^T P \right] \bar{e} + \left( w_u \bar{e}^T PB_c - \frac{1}{\rho^2} \bar{e}^T PB_c B_c^T P\bar{e} \right) \tag{8.60}$$

Observe that the term $w_u \bar{e}^T PB_c$ is scalar and can be written as $w_u \bar{e}^T PB_c = (1/2)\left( \bar{e}^T PB_c + B_c^T P\bar{e} \right) w_u$ then the last two terms of (8.60) can be written in the form

$$\left( w_u \bar{e}^T PB_c - \frac{1}{2\rho^2} \bar{e}^T PB_c B_c^T P\bar{e} \right) = -\frac{1}{2} \left( \frac{1}{\rho} B_c^T P\bar{e} - \rho w_u \right)^T \left( \frac{1}{\rho} B_c^T P\bar{e} - \rho w_u \right) + \frac{1}{2} \rho^2 w_u^2 \tag{8.61}$$

From (8.57) and (8.61), Equation 8.60 becomes

$$\dot{V} = -\frac{1}{2}\bar{e}^T Q\bar{e} - \frac{1}{2}\left(\frac{1}{\rho}B_c^T P\bar{e} - \rho w_u\right)^T \left(\frac{1}{\rho}B_c^T P\bar{e} - \rho w_u\right) + \frac{1}{2}\rho^2 w_u^2 \qquad (8.62)$$

which reduces to

$$\dot{V} \le -\frac{1}{2}\bar{e}^T Q\bar{e} + \frac{1}{2}\rho^2 w_u^2 \qquad (8.63)$$

Integrating both sides of (8.63) from 0 to $t$, we get

$$V(t) - V(0) \le -\frac{1}{2}\int_0^t \bar{e}^T Q\bar{e}\,d\tau + \frac{\rho^2}{2}\int_0^t w_u^2\,d\tau \qquad (8.64)$$

Since $V(t)$ is positive, then (8.64) implies (8.56). This means that the $H_\infty$ tracking performance index is satisfied. Moreover, it can be shown that the tracking error and the fuzzy approximation error are bounded by writing (8.64) as

$$V\left(\bar{e}(t), \psi(t), t\right) - V\left(\bar{e}(0), \psi(0), 0\right) \le \frac{1}{2}\int_0^t \rho^2 w_u^2\,d\tau < \infty \qquad (8.65)$$

Equation 8.65 implies that $\bar{e}(t)$ and $\psi(t)$ are bounded for $0 \le t < \infty$. This concludes the proof.

**Example 8.1**

Consider a third-order plant given by the transfer function

$$G(s) = \frac{s + 0.5}{s^3 + 2s^2 + 5s + 1}$$

Note that the system can be described in state space form (8.20) with

$$A = \begin{bmatrix} 0 & 1 & 0 \\ 0 & 0 & 1 \\ -1 & -5 & -2 \end{bmatrix}, \quad B = \begin{bmatrix} 0 \\ 0 \\ 1 \end{bmatrix}, \quad \text{and} \quad C = \begin{bmatrix} 0.5 & 1 & 0 \end{bmatrix}$$

Design a direct adaptive fuzzy controller in the form (8.25), (8.52), and (8.55) along with the adaptation law (8.42), such that the system output tracks the reference signal $y_r = \sin(t) + \sin(0.5t)$. It is clear that the system has relative degree $(n - m) = 3 - 1 = 2$, then the error dynamic equation takes the form (8.30) with

$$\Lambda_c = \begin{bmatrix} 0 & 1 \\ -k_0 & -k_1 \end{bmatrix}, \quad B_c = \begin{bmatrix} 0 \\ 1 \end{bmatrix}, \quad \text{and} \quad c_m = c_1 = 1$$

**Procedure:**

1. Choose the element of the vector $K^T = [k_0 \quad k_1] = [0.75 \quad 2.0]$.
2. Assume $Q = I$, $\sigma = 0.8$, and $\rho = 0.9$. Solve the Riccati-like equation
   (8.57) to get the matrix $P = \begin{bmatrix} 1.53 & 0.47 \\ 0.47 & 0.43 \end{bmatrix}$.
3. In this example, we assume there is no available human control
   experience in the form of IF-THEN rules (8.22).
4. Select a number of fuzzy sets for each state variable with
   Gaussian membership functions of the form $\mu_{F_i}^{\ell_i}(x_i) = e^{-\left(\frac{x_i - \bar{x}_i^{\ell_i}}{\sigma_i^{\ell_i}}\right)^2}$,
   where $\bar{x}_i^{\ell_i} = (h_i + q_i) + q_i(\ell_i - 1)$ with $h_i = -1$, $q_i = 0.33$, $\sigma_i^{\ell_i} = 0.05$,
   $i = 1, 2, 3$ and $\ell_i = 1, \dots, 5$.
5. The fuzzy basis functions are determined using (8.26).
6. The parameter vector $\theta$ is determined from (8.42) with design
   parameter $\gamma = 1500$.
7. Find the direct adaptive fuzzy control from (8.25).
8. Find the attenuation control term from (8.55).
9. The total control is determined from (8.52).

Simulation of the closed-loop system is carried out for the nominal system parameters. The system output and the reference signal are shown in Figure 8.1a and the total control signal is shown in Figure 8.1b.

The direct adaptive fuzzy controller can still achieve the desired tracking in the presence of parameter variation. To validate this feature, the system parameters represented by the last row of matrix $A$ are assumed to have $\pm 50\%$ mismatch of the nominal values. The system response and the control signal in these cases are shown in Figures 8.2 and 8.3.

## 8.5 Indirect Adaptive Fuzzy Tracking Controller

In the indirect adaptive fuzzy control, fuzzy system is used to model the unknown plant, and any available description of the plant behavior in the form of IF-THEN rules can be easily incorporated. To develop an indirect adaptive fuzzy controller, we consider the SISO linear plant given in (8.4) along with the matrices $A$, $B$, and $C$ defined after (8.4). The control objective is to design a controller $u_I(x, \theta)$ based on fuzzy logic system and an updating law for the adaptive parameter vector $\theta$ such that the plant output

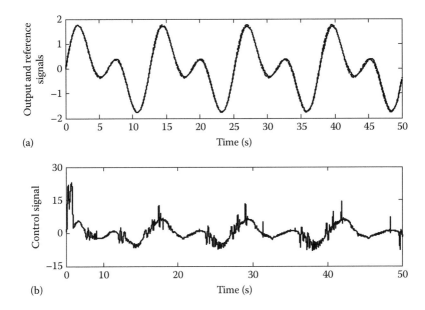

**FIGURE 8.1**
Tracking response: (a) system output (solid), reference (dashed), and (b) control signal for nominal system parameters.

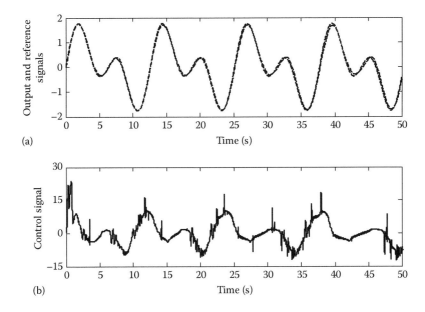

**FIGURE 8.2**
Tracking response: (a) system output (solid), reference (dashed), and (b) control signal for +50% mismatch in system parameters.

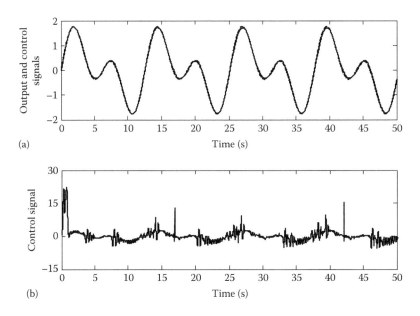

**FIGURE 8.3**
Tracking response: (a) system output (solid), reference (dashed), and (b) control signal for –50% mismatch in system parameters.

asymptotically tracks a given reference signal $y_r(t)$. When the system parameters are precisely known, the ideal controller takes the form given in (8.29) and repeated here:

$$u^* = \frac{1}{c_m}\left(y_r^{(n-m)} - \Phi(x) + K^T \bar{e}\right) \tag{8.66}$$

where $\Phi(x) = CA^{n-m}x$. In the case where the system parameters are unknown, the indirect adaptive fuzzy controller $u_I(x,\theta)$ is designed as

$$u_I(x,\theta) = \frac{1}{c_m}\left(y_r^{(n-m)} - \hat{\Phi}(x,\theta) + K^T \bar{e}\right) \tag{8.67}$$

where $\hat{\Phi}(x,\theta)$ is an adaptive fuzzy approximation of $\Phi(x)$ generated by the fuzzy logic system

$$\hat{\Phi}(x,\theta) = \frac{\sum_{\ell_1=1}^{m_1} \cdots \sum_{\ell_n=1}^{m_n} \bar{y}^{\ell_1 \cdots \ell_n}\left(\prod_{i=1}^{n} \mu_{F_i}^{\ell_i}(x_i)\right)}{\sum_{\ell_1=1}^{m_1} \cdots \sum_{\ell_n=1}^{m_n}\left(\prod_{i=1}^{n} \mu_{F_i}^{\ell_i}(x_i)\right)} \tag{8.68}$$

In order to attenuate the tracking error, an attenuation control term is used such that the overall control is given by

$$u = u_l(x, \theta) + u_a \tag{8.69}$$

Using (8.67) in (8.69), the error Equation 8.11 becomes

$$\dot{\bar{e}} = \Lambda \bar{e} + B_c \left[ \left( \hat{\Phi}(x, \theta) - \Phi(x) \right) - K^T \bar{e} - c_m u_a \right] \tag{8.70}$$

Define the minimum fuzzy approximation error $w_f$ as

$$w_f = \hat{\Phi}(x, \theta^*) - \Phi(x) \tag{8.71}$$

where $\theta^*$ is defined by

$$\theta^* = \arg\min_{\theta \in \Re^N} \left[ \sup_{x \in \Re^n} | \Phi(x) - \hat{\Phi}(x, \theta) | \right] \tag{8.72}$$

By adding and subtracting the term $\hat{\Phi}(x, \theta^*)$ to the right-hand side of (8.70), we get

$$\dot{\bar{e}} = \left( \Lambda - B_c K^T \right) \bar{e} + B_c \left[ \left( \hat{\Phi}(x, \theta) - \hat{\Phi}(x, \theta^*) \right) \right] + B_c \left[ \left( \hat{\Phi}(x, \theta^*) - \Phi(x) \right) \right] - c_m B_c u_a \tag{8.73}$$

Using Equation 8.71, the definition $\Lambda_c = (\Lambda - B_c K^T)$ and the representation of the fuzzy system $\hat{\Phi}(x, \theta) = \theta^T f_b$ in (8.73) leads to

$$\dot{\bar{e}} = \Lambda_c \bar{e} + B_c \psi^T f_b + B_c w_f - c_m B_c u_a \tag{8.74}$$

where $\psi^T = (\theta - \theta^*)^T$ is the parameter error vector.

The Lyapunov synthesis approach used in the foregoing direct adaptive fuzzy control can be repeated here for stability analysis and determining the adaptation law of the indirect adaptive fuzzy control. Choose the Lyapunov function (8.35), and find its time derivative as in (8.37). Evaluating this derivative along the trajectory (8.74) leads to

$$\dot{V} = \frac{1}{2} \bar{e}^T \left( \Lambda_c^T P + P \Lambda_c \right) \bar{e} + \psi^T \left( f_b \bar{e}^T P B_c + \frac{1}{\gamma} \dot{\theta} \right) + \left( w_f - c_m u_a \right) \bar{e}^T P B_c \tag{8.75}$$

The parameter updating law is selected as

$$\dot{\theta} = -\gamma f_b \bar{e}^T P B_c \tag{8.76}$$

Then

$$\dot{V} = \frac{1}{2}\overline{e}^T\left(\Lambda_c^T P + P\Lambda_c\right)\overline{e} + \left(w_f - c_m u_a\right)\overline{e}^T P B_c \tag{8.77}$$

If the attenuation control term is selected as (8.55), then following the same procedure given in the previous section, we can write (8.77) as

$$\dot{V} \le -\frac{1}{2}\overline{e}^T Q\overline{e} + \frac{1}{2}\rho^2 w_f^2 \tag{8.78}$$

where $Q$ is defined in (8.57). Therefore, with the adaptation law (8.76) and the attenuation control part (8.55), the same conclusion as that of Theorem 8.1 can be drawn.

**Example 8.2**

Consider the system defined by

$$G(s) = \frac{c_2 s^2 + c_0}{s^4 + a_3 s^3 + a_2 s^2 + a_1 s + a_0}$$

where $c_0 = 2, c_2 = 1, a_0 = 4, a_1 = 5, a_2 = 6$, and $a_3$ is unknown. The output of this system is required to track the reference signal $y_r = \cos(5t) + \sin(10t)$. The steps to design an indirect adaptive fuzzy controller are outlined in the following procedure:

1. Choose the element of vector $K^T = [k_0 \ k_1] = [\,0.5 \ \ 1.5\,]$.
2. Solve the Riccati-like equation (8.57) to get the matrix $P$ for a given $Q$ matrix. In the example, we assume $Q = 4000I$, $\sigma = 0.8$, and $\rho = 0.9$, and matrix $P$ is obtained as $P = \begin{bmatrix} 4055.8 & 55.8 \\ 55.8 & 55.8 \end{bmatrix}$.
3. Select five fuzzy sets for each state variable with Gaussian membership functions of the form $\mu_{F_i}^{\ell i}(x_i) = e^{-\left(\frac{x_i - \bar{x}_i^{\ell i}}{\sigma_i^{\ell i}}\right)^2}$.
4. The fuzzy basis functions are determined using (8.26).
5. The parameter vector $\theta$ is determined from (8.76) with the design parameter $\gamma = 250$.
6. Find the indirect adaptive fuzzy control from (8.67).
7. Find the attenuation control term from (8.55).
8. The total control is determined from (8.69).

The closed-loop output tracking response and the indirect adaptive control signal for three different values of the unknown parameter $a_3$ are obtained. The responses shown in Figures 8.4 through 8.6 are for the cases when $a_3 = 7, 14$, and 0, respectively.

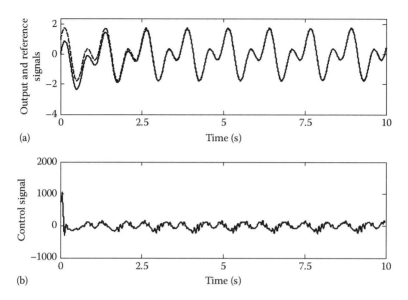

**FIGURE 8.4**
Tracking response: (a) system output (solid), reference (dashed), and (b) control signal for $a_3 = 7$.

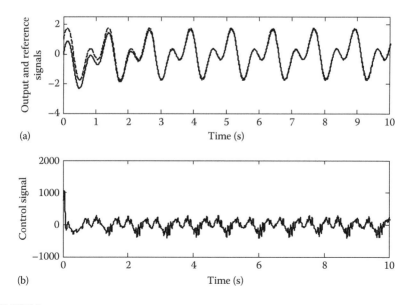

**FIGURE 8.5**
Tracking response: (a) system output (solid), reference (dashed), and (b) control signal for $a_3 = 14$.

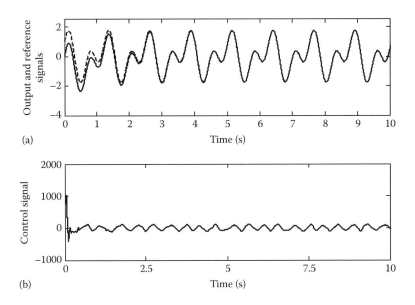

**FIGURE 8.6**
Tracking response: (a) system output (solid), reference (dashed), and (b) control signal for $a_3 = 0$.

## References

1. S. Sastry and M. Bodson, *Adaptive Control: Stability, Convergence, and Robustness*, Prentice-Hall, Englewood Cliffs, NJ, 1989.
2. H. J. Marques, *Nonlinear Control Systems: Analysis and Design*, John Wiley & Sons, Inc., Hoboken, NJ, 2003.
3. H. Khalil, *Nonlinear Systems*, 3rd edn., Prentice Hall, Upper Saddle River, NJ, 2002.
4. Y. C. Chang, Adaptive fuzzy-based tracking control for nonlinear SISO systems via VSS and H∞ approaches, *IEEE Trans. Fuzzy Syst.*, 9(2), 278–292, April 2001.

# 9

# *Direct Adaptive Fuzzy Load Frequency Control*

## 9.1 Introduction

The operation of power systems dictates matching the power generation and the load power. The role of load frequency control (LFC) is to ensure the balance between the generation and load in a given area. Practically speaking, the LFC of power systems encounters parameter and model uncertainties. The classical control techniques of LFC which are based on mathematical models fail to achieve the control objectives in the presence of uncertainties or changes in the operating points. In this chapter, we utilize the direct adaptive fuzzy logic control (DAFLC) methodology developed in Chapter 8 to design an LFC scheme for a multi-area power system. The power system is assumed to encounter unknown parameter variation in its components.

## 9.2 Controller Canonical Form of a Multi-Area Power System

The model of a multi-area power system was developed in Chapter 2 as

$$
\left.
\begin{aligned}
\dot{\underline{x}}_i &= \underline{A}_i \underline{x}_i + \underline{B}_i u_i + \sum_{j=1, j \neq i}^{N} \underline{A}_{ij} \underline{x}_j + \underline{F}_i \Delta P_{d_i} \\
\underline{y}_i &= \underline{C}_i \underline{x}_i
\end{aligned}
\right\}
\tag{9.1}
$$

where $\underline{x}_i = \left[ \Delta f_i \, \Delta P_{mi} \, \Delta P_{gi} \, \Delta P_{tiei} \right]^T$ and the model matrices are given by

$$\underline{A}_i = \begin{bmatrix} -D_i/2H_i & 1/2H_i & 0 & -1/2H_i \\ 0 & -1/T_{t_i} & 1/T_{t_i} & 0 \\ -1/\left(R_iT_{g_i}\right) & 0 & -1/T_{g_i} & 0 \\ 2\pi\sum\limits_{j=1,\,j\neq i}^{N} T_{ij} & 0 & 0 & 0 \end{bmatrix} \tag{9.2}$$

$$\underline{A}_{ij} = \begin{bmatrix} 0 & 0 & 0 & 0 \\ 0 & 0 & 0 & 0 \\ 0 & 0 & 0 & 0 \\ -2\pi T_{ij} & 0 & 0 & 0 \end{bmatrix} \tag{9.3}$$

$$\underline{B}_i^T = \begin{bmatrix} 0 & 0 & \dfrac{1}{T_{g_i}} & 0 \end{bmatrix} \tag{9.4}$$

$$\underline{F}_i^T = \begin{bmatrix} -\dfrac{1}{2H_i} & 0 & 0 & 0 \end{bmatrix} \tag{9.5}$$

$$\underline{C}_i = \begin{bmatrix} 1 & 0 & 0 & 0 \end{bmatrix} \tag{9.6}$$

To facilitate the design of a direct adaptive fuzzy controller for each area, the model (9.1) is transformed to the controller canonical form. A given state model can be transformed to the controller canonical form via coordinate transformation. The coordinate transformation exists if the state model is controllable. This concept is presented next.

### 9.2.1 Coordinate Transformation

For an $n$-dimensional linear time-invariant controllable state model

$$\left.\begin{array}{l} \dot{x} = Ax + Bu \\ y = Cx \end{array}\right\} \tag{9.7}$$

Any *nonsingular* $n \times n$ matrix $T$ defines a coordinate transformation through the relation

$$x = Tz \tag{9.8}$$

where $z$ is the transformed state. The transformed state model is given by

$$\left. \begin{aligned} \dot{z} &= \hat{A}z + \hat{B}u \\ y &= \hat{C}z \end{aligned} \right\} \tag{9.9}$$

where

$\hat{A} = T^{-1}AT$
$\hat{B} = T^{-1}B$
$\hat{C} = CT$

The state coordinate transformation (9.8) permits the construction of special state-space realizations that facilitate a particular type of analysis and controller design. As shown in Chapter 8, the controller canonical form makes the design of a tracking controller straightforward. Suppose the controller canonical form of the state model (9.7) is given by

$$\left. \begin{aligned} \dot{x}_{cc} &= A_{cc}x_{cc} + B_{cc}u_{cc} \\ y &= C_{cc}x_{cc} \end{aligned} \right\} \tag{9.10}$$

in which

$$A_{cc} = \begin{bmatrix} 0 & 1 & 0 & \cdots & 0 \\ 0 & 0 & 1 & \cdots & 0 \\ \vdots & \vdots & \vdots & \ddots & \vdots \\ 0 & 0 & 0 & 0 & 1 \\ -a_0 & -a_1 & -a_2 & \cdots & -a_{n-1} \end{bmatrix}, \ B_{cc} = \begin{bmatrix} 0 \\ 0 \\ \vdots \\ \vdots \\ 1 \end{bmatrix}, \ \text{and} \ C_{cc} = \begin{bmatrix} c_0 & c_1 & \cdots & c_m & 0 \cdots 0 \end{bmatrix}$$

The form (9.10) can be obtained by the state coordinate transformation

$$x = T_{cc}x_{cc} \tag{9.11}$$

The state coordinate transformation (9.11) preserves the controllability of the actual model (9.7) and the transformed model (9.10) [1]. A system of the form (9.7) is said to be controllable if the controllability matrix given by

$$P = \begin{bmatrix} B & AB \ldots A^{n-1}B \end{bmatrix} \tag{9.12}$$

has a full rank. To show that the controllability is invariant under similarity transformation, we find the controllability matrix of (9.10) in terms of the controllability matrix of (9.7). The controllability matrix $P_{cc}$ of (9.10) is given by

$$P_{cc} = \left[ B_{cc} \quad A_{cc}B_{cc} \ldots A_{cc}^{n-1}B_{cc} \right] \tag{9.13}$$

Using the forms $A_{cc} = T_{cc}^{-1}AT_{cc}$ and $B_{cc} = T_{cc}^{-1}B$ in (9.13), we get

$$
\begin{aligned}
P_{cc} &= \left[ B_{cc} \quad A_{cc}B_{cc} \ldots A_{cc}^{n-1}B_{cc} \right] \\
&= \left[ T_{cc}^{-1}B \quad T_{cc}^{-1}AT_{cc}\left(T_{cc}^{-1}B\right) \ldots T_{cc}^{-1}A^{n-1}T_{cc}\left(T_{cc}^{-1}B\right) \right] \\
&= \left[ T_{cc}^{-1}B \quad T_{cc}^{-1}AB \ldots T_{cc}^{-1}A^{n-1}B \right] \\
&= T_{cc}^{-1}\left[ B \quad AB \ldots A^{n-1}B \right] \\
&= T_{cc}^{-1}P
\end{aligned}
\tag{9.14}
$$

Since $T_{cc}$ is nonsingular matrix

$$\text{rank}\left(P_{cc}\right) = \text{rank}\left(P\right) \tag{9.15}$$

Hence, the controllability is invariant under state transformation.

**Lemma 9.1 [2]**

The controllability matrix of the $n$-dimensional controller canonical form (9.10) is given by

$$
P_{cc} = \begin{bmatrix}
a_1 & a_2 & \cdots & a_{n-1} & 1 \\
a_2 & a_3 & \cdots & 1 & 0 \\
\vdots & \vdots & \ddots & \vdots & \vdots \\
a_{n-1} & 1 & \cdots & 0 & 0 \\
1 & 0 & \cdots & 0 & 0
\end{bmatrix}^{-1}
\tag{9.16}
$$

where $a_i$, $i=1,\ldots,n-1$ are the coefficients of the characteristic polynomial

$$\alpha(s) = |sI - A| = s^n + a_{n-1}s^{n-1} + \cdots + a_1 s + a_0 \qquad (9.17)$$

The transformation matrix $T_{cc}$ is determined from (9.14) as

$$T_{cc} = \mathrm{PP}_{cc}^{-1} \qquad (9.18)$$

## 9.2.2 Controller Canonical Form of a Multi-Area LFC System

The multi-area LFC model given by (9.1) can be transformed to the controller canonical form as follows. Let $x_i$ be the transformed state where

$$\underline{x}_i = T_i x_i \qquad (9.19)$$

$$x_i = T_i^{-1}\underline{x}_i \qquad (9.20)$$

Substituting (9.1) in the differentiation of (9.20), we get

$$\dot{x}_i = T_i^{-1}\dot{\underline{x}}_i = T_i^{-1}\underline{A}_i \underline{x}_i + T_i^{-1}\underline{B}_i u_i + T_i^{-1}\sum_{j=1,j\neq i}^{N} \underline{A}_{ij}\underline{x}_j + T_i^{-1}\underline{F}_i \Delta P_{d_i} \qquad (9.21)$$

Using (9.19) in (9.21) yields

$$\left.\begin{array}{l} \dot{x}_i = A_i x_i + B_i u_i + \displaystyle\sum_{j=1,j\neq i}^{N} A_{ij}x_j + F_i \Delta P_{d_i} \\[2em] y_i = C_i x_i \end{array}\right\} \qquad (9.22)$$

where

$A_i = T_i^{-1}\underline{A}_i T_i$

$A_{ij} = T_i^{-1}\underline{A}_{ij}T_j$

$B_i = T_i^{-1}\underline{B}_i$

$F_i = T_i^{-1}\underline{F}_i$

$C_i = \underline{C}_i T_i$

The transformation matrix is determined from (9.18) as $T_i = \underline{P}_i P_i^{-1}$ where $\underline{P}_i$ and $P_i$ are the controllability matrices of (9.1) and (9.22), respectively.

**Example 9.1**

The two-area system given in Example 2.5 is considered here. The system parameters are given in Table 9.1.

The matrices of each area are calculated from (9.2) through (9.6) as

$$
\underline{A}_1 = \begin{bmatrix} -0.06 & 0.1 & 0 & -0.1 \\ 0 & -2 & 2 & 0 \\ -100 & 0 & -5 & 0 \\ 2 & 0 & 0 & 0 \end{bmatrix}, \quad \underline{A}_2 = \begin{bmatrix} -0.1125 & 0.125 & 0 & -0.125 \\ 0 & -1.67 & 1.67 & 0 \\ -53.33 & 0 & -3.33 & 0 \\ 2 & 0 & 0 & 0 \end{bmatrix},
$$

$$
\underline{B}_1 = \begin{bmatrix} 0 \\ 0 \\ 5 \\ 0 \end{bmatrix}, \quad \underline{B}_2 = \begin{bmatrix} 0 \\ 0 \\ 3.33 \\ 0 \end{bmatrix}, \quad \underline{F}_1 = \begin{bmatrix} -0.1 \\ 0 \\ 0 \\ 0 \end{bmatrix}, \quad \underline{F}_2 = \begin{bmatrix} -0.125 \\ 0 \\ 0 \\ 0 \end{bmatrix}, \quad \text{and } \underline{C}_1 = \underline{C}_2 = \begin{bmatrix} 1 & 0 & 0 & 0 \end{bmatrix},
$$

The interconnection matrices are determined as $\underline{A}_{12} = \underline{A}_{21} = \begin{bmatrix} 0 & 0 & 0 & 0 \\ 0 & 0 & 0 & 0 \\ 0 & 0 & 0 & 0 \\ -2 & 0 & 0 & 0 \end{bmatrix}$.

It is required to transform this system representation to a controller canonical form.

**Solution:**

The controllability matrix of each area is calculated from $\underline{P}_i = \begin{bmatrix} \underline{B}_i & \underline{A}_i \underline{B}_i & \underline{A}_i^2 \underline{B}_i & \underline{A}_i^3 \underline{B}_i \end{bmatrix}$ and found as

$$
\underline{P}_1 = \begin{bmatrix} 0 & 0 & 1 & -7.06 \\ 0 & 10 & -70 & 390 \\ 5 & -25 & 125 & -725 \\ 0 & 0 & 0 & 2 \end{bmatrix}
$$

$$
\underline{P}_2 = \begin{bmatrix} 0 & 0 & 0.69 & -3.55 \\ 0 & 5.55 & -27.77 & 108.02 \\ 3.33 & -11.11 & 37.04 & -160.5 \\ 0 & 0 & 0 & 1.39 \end{bmatrix}
$$

**TABLE 9.1**

Parameters of a Two-Area System

| Area #1 | | | | | | Area #2 | | | | | |
|---|---|---|---|---|---|---|---|---|---|---|---|
| $T_{g1}$ | $T_{t1}$ | $H_1$ | $D_1$ | $R_1$ | $\beta_1$ | $T_{g2}$ | $T_{t2}$ | $H_2$ | $D_2$ | $R_2$ | $\beta_2$ |
| 0.2 | 0.5 | 5 | 0.6 | 0.05 | 20.6 | 0.3 | 0.6 | 4 | 0.9 | 0.0625 | 16.9 |

To find the transformation matrices $T_1$ and $T_2$, first we have to determine the controllability matrices of the controller form $P_1$ and $P_2$ in terms of the characteristic equation coefficients. The characteristic equations of area 1 and area 2 are found from (9.17) as

$$\alpha_1(s) = s^4 + 7.06s^3 + 10.62s^2 + 22s + 2$$

$$\alpha_2(s) = s^4 + 5.11s^3 + 6.37s^2 + 12.98s + 1.39$$

Using the coefficients of these polynomials in (9.16) yields

$$P_1 = \begin{bmatrix} 0 & 0 & 0 & 1 \\ 0 & 0 & 1 & -7.06 \\ 0 & 1 & -7.06 & 39.22 \\ 1 & -7.06 & 39.22 & -223.94 \end{bmatrix}$$

$$P_2 = \begin{bmatrix} 0 & 0 & 0 & 1 \\ 0 & 0 & 1 & -5.11 \\ 0 & 1 & -5.11 & 19.77 \\ 1 & -5.11 & 19.77 & -81.5 \end{bmatrix}$$

Then, the transformation matrices are calculated from (9.18) as

$$T_1 = \underline{P}_1 P_1^{-1} = \begin{bmatrix} 0 & 1 & 0 & 0 \\ 2 & 0.6 & 10 & 0 \\ 2 & 1.6 & 10.3 & 5 \\ 2 & 0 & 0 & 0 \end{bmatrix}$$

$$T_2 = \underline{P}_2 P_2^{-1} = \begin{bmatrix} 0 & 0.69 & 0 & 0 \\ 1.39 & 0.62 & 5.55 & 0 \\ 1.39 & 1.45 & 5.93 & 3.33 \\ 1.39 & 0 & 0 & 0 \end{bmatrix}$$

The following matrices of the controller canonical form for each area are obtained:

*Area 1:*

$$A_1 = T_1^{-1} \underline{A}_1 T_1 = \begin{bmatrix} 0 & 1 & 0 & 0 \\ 0 & 0 & 1 & 0 \\ 0 & 0 & 0 & 1 \\ -2 & -22 & -10.62 & -7.06 \end{bmatrix}, \quad B_1 = T_1^{-1} \underline{B}_1 = \begin{bmatrix} 0 \\ 0 \\ 0 \\ 1 \end{bmatrix},$$

$$F_1 = T_1^{-1} \underline{F}_1 = \begin{bmatrix} 0 \\ -0.1 \\ 0.006 \\ 0.019 \end{bmatrix}, \quad \text{and} \quad C_1 = \underline{C}_1 T_1 = \begin{bmatrix} 0 & 1 & 0 & 0 \end{bmatrix}.$$

*Area 2:*

$$A_2 = T_2^{-1}\underline{A_2}T_2 = \begin{bmatrix} 0 & 1 & 0 & 0 \\ 0 & 0 & 1 & 0 \\ 0 & 0 & 0 & 1 \\ -1.39 & -12.98 & -6.37 & -5.11 \end{bmatrix}, \quad B_2 = T_2^{-1}\underline{B_2} = \begin{bmatrix} 0 \\ 0 \\ 0 \\ 1 \end{bmatrix}$$

$$F_2 = T_2^{-1}\underline{F_2} = \begin{bmatrix} 0 \\ -0.18 \\ 0.02 \\ 0.043 \end{bmatrix}, \quad \text{and} \quad C_2 = \underline{C_2}T_2 = \begin{bmatrix} 0 & 0.69 & 0 & 0 \end{bmatrix}.$$

Interconnection matrices:

$$A_{12} = T_1^{-1}\underline{A_{12}}T_2 = \begin{bmatrix} 0 & -0.69 & 0 & 0 \\ 0 & 0 & 0 & 0 \\ 0 & 0.14 & 0 & 0 \\ 0 & -0.008 & 0 & 0 \end{bmatrix} \quad \text{and} \quad A_{21} = T_2^{-1}\underline{A_{21}}T_1 = \begin{bmatrix} 0 & -01.44 & 0 & 0 \\ 0 & 0 & 0 & 0 \\ 0 & 0.36 & 0 & 0 \\ 0 & -0.04 & 0 & 0 \end{bmatrix}$$

## 9.3 Design of DAFLC Load Frequency Control for a Multi-Area Power System

In this section, we follow the methodology presented in Chapter 8 to develop an LFC scheme based on the DAFLC for an $N$-area power system. It is assumed that the multi-area system is written in controller canonical form (9.19 and 9.20). The isolated and undisturbed power area model is written from (9.22) as

$$\left. \begin{array}{l} x_i = A_i x_i + B_i u_i \\ y_i = C_i x_i \end{array} \right\} \tag{9.23}$$

where

$$A_i = \begin{bmatrix} 0 & 1 & 0 & 0 \\ 0 & 0 & 1 & 0 \\ 0 & 0 & 0 & 1 \\ -a_{0i} & -a_{1i} & -a_{2i} & -a_{3i} \end{bmatrix}$$

$B_i = [0 \ \ 0 \ \ 0 \ \ 1]^T$

$C_i = [0 \ \ c_{1i} \ \ 0 \ \ 0]$

The transfer function realization of (9.23) is given by

$$G(s) = \frac{c_{1i}s}{s^4 + a_{3i}s^3 + a_{2i}s^2 + a_{1i}s + a_{0i}}$$

This indicates that the relative degree $r_i = n_i - m_i = 4 - 1 = 3$. When the parameters of (9.23) are fully known, the ideal tracking controller $u_i^*$ (similar to (8.29)) can be written as

$$u_i^* = \frac{1}{c_{1i}}\left(\ddot{y}_{d_i} - C_i A_i^3 x_i + K_i^T \bar{e}_i\right) \qquad (9.24)$$

where

$$\bar{e}_i = \left[\left(y_{d_i} - y_i\right) \quad \cdots \quad \left(\ddot{y}_{d_i} - \ddot{y}_i\right)\right]^T$$

$y_{d_i}$ is the desired output signal

The vector $K_i^T = \left[k_{0i} \quad k_{1i} \quad k_{2i}\right]$, $i = 1, \dots N$ is chosen such that the characteristic polynomial $\lambda_i(s) = s^3 + k_{2i}s^2 + k_{1i}s + k_{0i}$ has all roots in the open left-hand side of the s-plane.

The interconnected area subjected to constant load disturbance can be written in the form

$$\left.\begin{aligned}\dot{x}_i &= A_i x_i + B_i u_i + M_i(X_i) + D_i \\ y_i &= C_i x_i\end{aligned}\right\} \qquad (9.25)$$

where

$$M_i(X_i) = \sum_{j=1, j \neq i}^{N} A_{ij} x_j = \left(A_{i1}x_1 + A_{i2}x_2 + \cdots A_{iN}x_N\right) = \begin{bmatrix} m_{11_i}(X_i) \\ m_{21_i}(X_i) \\ m_{31_i}(X_i) \\ m_{41_i}(X_i) \end{bmatrix}$$

$$D_i = F_i \Delta P_{d_i} = \begin{bmatrix} d_{11_i} & d_{21_i} & d_{31_i} & d_{41_i} \end{bmatrix}^T$$

and $X_i = \left[x_1 \quad x_2 \cdots x_j \cdots x_N\right]_{j \neq i}^T$ is the composite state vector of the areas connected to the $i$th area. The third derivative of the output can be written from (9.25) as

$$\dddot{y}_i = C_i A_i^3 x_i + C_i A_i^2 B_i u_i + \left(C_i A_i^2 M_i + C_i A_i \dot{M}_i + C_i \ddot{M}_i\right) + C_i A_i^2 D_i \qquad (9.26)$$

Note that $C_i A_i^2 B_i = c_{1i}$, $C_i A_i^2 M_i = c_{1i} m_{41_i}$, $C_i A_i \dot{M}_i = c_{1i} \dot{m}_{31_i}$, $C_i \ddot{M}_i = c_{1i} \ddot{m}_{21_i}$ and $C_i A_i^2 D_i = c_{1i} d_{41_i}$.

Then, the preceding equation becomes

$$\dddot{y}_i = \Gamma(x_i) + c_{1i} u_i + c_{1i} \chi_i(X_i) + c_{1i} d_{41_i} \tag{9.27}$$

where $\Gamma(x_i) = C_i A_i^3 x_i$ and $\chi_i(X_i) = (m_{41_i} + \dot{m}_{31_i} + \ddot{m}_{21_i})$, which accounts for the interconnection terms. When the system parameters are unknown, the direct adaptive fuzzy control $u_{di}(x_i, \theta_i)$ defined in (8.23) (or equivalently (8.25)) is employed in (9.27). Moreover, the interconnection term $\chi_i(X_i)$ is assumed to be unknown and hence can be approximated using fuzzy logic system as $\hat{\chi}_i(X_i, \phi_i)$. Therefore, the control law of each interconnected and disturbed area $u_i$ is determined as

$$u_i = u_{di}(x_i, \theta_i) + u_{ai} - \hat{\chi}_i(X_i, \phi_i) \tag{9.28}$$

where $\theta_i$ and $\phi_i$ are adaptive parameters whose updating laws will be determined later. The control term $u_{ai}$ is used to attenuate the tracking error in the sense of $H_\infty$ performance.

Using the preceding equation in (9.27), we get

$$\dddot{y}_i = C_i A_i^3 x_i + c_{1i} \left( u_{di}(x_i, \theta_i) + u_{ai} + \chi_i(X_i) - \hat{\chi}_i(X_i, \phi_i) \right) + c_{1i} d_{41_i} \tag{9.29}$$

The terms $u_{di}(x_i, \theta_i)$ and $\hat{\chi}_i(X_i, \phi_i)$ are defined in terms of the fuzzy basis functions $f_{bi}(x_i)$ and $g_{bi}(X_i)$ as

$$u_{di}(x_i, \theta_i) = \theta_i^T f_{bi}(x_i) \tag{9.30}$$

$$\hat{\chi}_i(X_i, \phi_i) = \phi_i^T g_{bi}(X_i) \tag{9.31}$$

The control law (9.28) consists of three terms, the direct adaptive fuzzy control $u_{di}(x_i, \theta_i)$, the attenuation control part $u_{ai}$, and the approximation of the interconnection terms $\hat{\chi}_i(X_i, \phi_i)$. Therefore, the control law is termed as a combined direct–indirect adaptive fuzzy controller in which the term $u_{di}(x_i, \theta_i)$ represents the direct control part and the term $\hat{\chi}_i(X_i, \phi_i)$ represents the indirect control part. Rearrange (9.24) as

$$\dddot{y}_{d_i} = C_i A_i^3 x_i - K_i^T \bar{e}_i + c_{1i} u_i^* \tag{9.32}$$

and then using it in (9.29) leads to the following error equation:

$$\ddot{e}_i = -K_i^T \overline{e}_i + c_{1i}\left(u_i^* - u_{di}\left(x_i,\theta_i\right)\right) + c_{1i}\left(\hat{\chi}_i\left(X_i,\phi_i\right) - \chi_i\left(X_i\right)\right) - c_{1i}u_{ai} - c_{1i}d_{41i} \quad (9.33)$$

This equation can be put in the following state space form:

$$\dot{\overline{e}}_i = \Lambda_{ci}\overline{e}_i + c_{1i}B_{ci}\left[\left(u_i^* - u_{di}\left(x_i,\theta_i\right)\right) + \hat{\chi}_i\left(X_i,\phi_i\right) - \chi_i\left(X_i\right)\right) - u_{ai} - d_{41i}\right] \quad (9.34)$$

where

$$\Lambda_{ci} = \begin{bmatrix} 0 & 1 & 0 \\ 0 & 0 & 1 \\ -k_{0i} & -k_{2i} & -k_{2i} \end{bmatrix}$$

$$B_{ci} = [0 \ \ 0 \ \ 1]^T$$

### 9.3.1 Parameter Updating Laws

Define the mismatch between the ideal control and direct adaptive fuzzy control as

$$w_{ui} = u_i^* - u_{di}\left(x_i,\theta_i\right)$$

and the mismatch between fuzzy approximation of the interconnection term and its actual value as

$$w_{inti} = \hat{\chi}_i\left(X_i,\phi_i\right) - \chi_i\left(X_i\right)$$

Suppose the optimum values of the adaptive parameters $\theta_i^*$ and $\phi_i^*$ are defined by

$$\left.\begin{aligned} \theta_i^* &= \arg\min_{\theta_i}\left[\sup|w_{ui}|\right] \\ \phi_i^* &= \arg\min_{\phi_i}\left[\sup|w_{inti}|\right] \end{aligned}\right\} \quad (9.35)$$

Then the mismatches $w_{ui}$ and $w_{inti}$ will attain the following minimum:

$$w_{ui\min} = u_i^* - u_{di}\left(x_i,\theta_i^*\right) \quad (9.36)$$

$$w_{inti\min} = \hat{\chi}_i\left(X_i,\phi_i^*\right) - \chi_i\left(X_i\right) \quad (9.37)$$

Adding and subtracting the terms $u_{di}\left(x_i,\theta_i^*\right)$ and $\hat{\chi}_i\left(X_i,\phi_i^*\right)$ to (9.34) and then using (9.36) and (9.37) leads to the following error equation:

$$\dot{\bar{e}}_i = \Lambda_{ci}\bar{e}_i + c_{1i}B_{ci}\left(\Delta u_i - \Delta \text{int}_i - u_{ai} - d_{41_i} + w_i\right) \tag{9.38}$$

where

$\Delta u_i = u_{di}\left(x_i,\theta_i^*\right) - u_{di}\left(x_i,\theta_i\right)$

$\Delta \text{int}_i = \hat{\chi}_i\left(X_i,\phi_i^*\right) - \hat{\chi}_i\left(X_i,\phi_i\right)$

$w_i = w_{uimin} + w_{intimin}$

Using (9.30) and (9.31), we can write $\Delta u_i$ and $\Delta \text{int}_i$ as

$$\Delta u_i = \left(\theta_i^* - \theta_i\right)^T f_{bi} \tag{9.39}$$

$$\Delta \text{int}_i = \left(\phi_i^* - \phi_i\right)^T g_{bi} \tag{9.40}$$

In order to determine the updating laws of the parameter vectors $\theta_i$ and $\phi_i$, we use the Lyapunov synthesis approach. Consider a positive definite Lyapunov function defined by

$$V_i = \frac{1}{2}\bar{e}_i^T P_i \bar{e}_i + \frac{1}{2\gamma_{1i}}\psi_i^T \psi_i + \frac{1}{2\gamma_{2i}}\varphi_i^T \varphi_i \tag{9.41}$$

where $\psi_i = \theta_i^* - \theta_i$ and $\varphi_i = \phi_i^* - \phi_i$ are the parameter error vectors.
Taking the time derivative of both sides of (9.41) yields

$$\dot{V}_i = \frac{1}{2}\left(\dot{\bar{e}}_i^T P_i \bar{e}_i + \bar{e}_i^T P_i \dot{\bar{e}}_i\right) + \frac{1}{\gamma_{1i}}\psi_i^T \dot{\psi}_i + \frac{1}{\gamma_{2i}}\varphi_i^T \dot{\varphi}_i \tag{9.42}$$

Using (9.38 through 9.40) into (9.42), we get

$$\dot{V}_i = \frac{1}{2}\bar{e}_i^T\left(\Lambda_{ci}^T P_i + P_i \Lambda_{ci}\right)\bar{e}_i + \psi_i^T\left(c_{1i}B_{ci}^T P_i \bar{e}_i f_{bi} - \frac{1}{\gamma_{1i}}\dot{\theta}_i\right) - \varphi_i^T\left(c_{1i}B_{ci}^T P_i \bar{e}_i g_{bi} + \frac{1}{\gamma_{2i}}\dot{\phi}_i\right)$$
$$+ c_{1i}\left(-u_{ai} - d_{41_i} + w_i\right)B_{ci}^T P_i \bar{e}_i \tag{9.43}$$

If the parameter updating laws are chosen as

$$\dot{\theta}_i = \gamma_{1i} c_{1i} B_{ci}^T P_i \bar{e}_i f_{bi} \tag{9.44}$$

$$\dot{\phi}_i = -\gamma_{2i} c_{1i} B_{ci}^T P_i \bar{e}_i g_{bi} \tag{9.45}$$

then the second and third terms on the right-hand side of (9.43) are nullified. Now, we state the following theorem

**Theorem 9.1**

For each interconnected and disturbed LFC area given in (9.25), the direct-indirect adaptive fuzzy controller given by (9.28), (9.30), and (9.31) along with the parameter updating laws (9.44) and (9.45) ensures the boundedness of the tracking error and the parameter error. Moreover, the tracking error achieves the $H_\infty$ tracking performance if the attenuation control term is chosen as

$$u_{ai} = \frac{1}{c_{1i} \sigma_i} \bar{e}_i^T P_i B_{ci} \tag{9.46}$$

where
  $\sigma_i$ is a positive constant
  $P_i$ is a positive definite matrix solution of the following Riccati-like equation:

$$Q_i = -\left[ \left( \Lambda_{ci}^T P_i + P_i \Lambda_{ci} \right) - P_i B_{ci} \left( \frac{2}{\sigma_i} - \frac{1}{\rho_i^2} \right) B_{ci}^T P_i \right] \tag{9.47}$$

*Proof:*
Substituting (9.44 through 9.46) into (9.43) leads to

$$\dot{V}_i = \frac{1}{2} \bar{e}_i^T \left( \Lambda_{ci}^T P_i + P_i \Lambda_{ci} \right) \bar{e}_i - \frac{1}{\sigma_i} \bar{e}_i^T P_i B_{ci} B_{ci}^T P_i \bar{e}_i + c_{1i} \left( w_i - d_{41_i} \right) B_{ci}^T P_i \bar{e}_i \tag{9.48}$$

Adding and subtracting the term $\dfrac{1}{2\rho_i} \bar{e}_i^T P_i B_{ci} B_{ci}^T P_i \bar{e}_i$ in (9.48) and using (9.47), we get

$$\dot{V}_i = -\frac{1}{2} \bar{e}_i^T Q_i \bar{e}_i + w_{Ti} B_{ci}^T P_i \bar{e}_i - \frac{1}{2\rho_i^2} \bar{e}_i^T P_i B_{ci} B_{ci}^T P_i \bar{e}_i \tag{9.49}$$

where

$$w_{Ti} = c_{1i} \left( w_i - d_{41_i} \right) \tag{9.50}$$

Note that the last two terms on the right-hand side of (9.49) can be rewritten in the form

$$w_{Ti}B_{ci}^T P_i \bar{e}_i - \frac{1}{2\rho_i^2} \bar{e}_i^T P_i B_{ci} B_{ci}^T P_i \bar{e}_i = -\frac{1}{2}\left(\frac{1}{\rho_i} B_{ci}^T P_i \bar{e}_i - \rho_i w_{Ti}\right)^T$$

$$\times \left(\frac{1}{\rho_i} B_{ci}^T P_i \bar{e}_i - \rho_i w_{Ti}\right) + \frac{1}{2}\rho_i^2 w_{Ti}^2 \qquad (9.51)$$

Using (9.51) in (9.49), we write the following inequality:

$$\dot{V}_i \leq -\frac{1}{2}\bar{e}_i^T Q_i \bar{e}_i + \frac{1}{2}\rho_i^2 w_{Ti}^2 \qquad (9.52)$$

The same argument presented in Theorem 8.1 can be repeated here to conclude that both the tracking error and fuzzy approximation error are bounded and the tracking error achieves the $H_\infty$ tracking performance.

A block diagram of the combined direct–indirect adaptive fuzzy load frequency controller is shown in Figure 9.1.

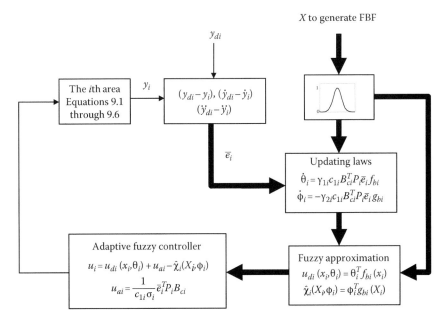

**FIGURE 9.1**
Block diagram of the proposed direct–indirect adaptive fuzzy controller for a multi-area power system.

**Example 9.2**

In this example, we apply the adaptive fuzzy controller given by (9.28 through 9.31) and (9.44 and 9.45) for the two-area system of Example 9.1. The matrices of each area are shown in Example 9.1. The procedure to design the controller is as follows:
*Off-line calculation*

1. Determine the state transformation matrices $T_1$ and $T_2$ to transform the interconnected two-area system to controller canonical form. These matrices are found in Example 9.1. Note that in the proposed controller, we need $c_{1i}$, which is model-dependent. It can be determined from the controller canonical form matrix $C_i$ or from the numerator coefficient of the local area transfer function. These values are found as $c_{11} = 1$ and $c_{12} = 0.69$.

2. Find the system matrices of the controller canonical form, as shown in Example 9.1.

3. Select suitable values of $K_i^T = \begin{bmatrix} k_{0i} & k_{1i} & k_{2i} \end{bmatrix}$ such that $\lambda_i(s) = s^3 + k_{2i}s^2 + k_{1i}s + k_{0i}$ is Hurwitz for $i = 1, 2$. The two vectors are chosen as $K_1^T = K_2^T = \begin{bmatrix} 0.04 & 0.44 & 0.7 \end{bmatrix}$.

4. Select a positive definite $3 \times 3$ matrix $Q_i$ and positive constants $\sigma_i$ and $\rho_i$ such that the term $\left( (2/\sigma_i) - (1/\rho_i^2) \right)$ is positive. These parameters are specified as $\sigma_1 = \sigma_2 = 1$, $\rho_1 = \rho_2 = 2$ and the matrices $Q_1 = Q_2 = 0.05I$.

5. Solve the Riccati-like equation (9.47) to determine the positive definite matrices $P_1$ and $P_2$ as $P_1 = P_2 \begin{bmatrix} 0.16 & 0.21 & 0.15 \\ 0.21 & 0.43 & 0.32 \\ 0.15 & 0.32 & 0.35 \end{bmatrix}$.

6. Define the number of fuzzy sets for each state variable. Here, we select five Gaussian membership functions for each state variable $x_i$, $i = 1, 2, \ldots, 8$. The membership functions of the frequency, mechanical power, governor power, and tie-line power deviations for area 1 and area 2 are chosen similar and shown in Figure 9.2.

7. Specify two fuzzy basis functions $f_{bi}(x_i)$ and $g_{bi}(X_i)$ given in (9.30) and (9.31), where $x_1 = [x_{11}\, x_{12}\, x_{13}\, x_{14}]^T$ and $x_2 = [x_{21}\, x_{22}\, x_{23}\, x_{24}]^T$ are the state vectors of each area and $X_i$ is the state vector of area 2 when $i = 1$ and of area 1 when $i = 2$. The form of each element of the fuzzy basis functions is given by

$$f_{bi}^k(x_i) = \frac{\prod_{r=1}^{4} \mu_{F_r}^{\ell_r}(x_{ir})}{\sum_{\ell_1=1}^{5} \sum_{\ell_2=1}^{5} \sum_{\ell_3=1}^{5} \sum_{\ell_4=1}^{5} \sum_{r=1}^{4} \mu_{F_r}^{\ell_r}(x_{ir})}, \quad k = 1, 2 \ldots, 5^4$$

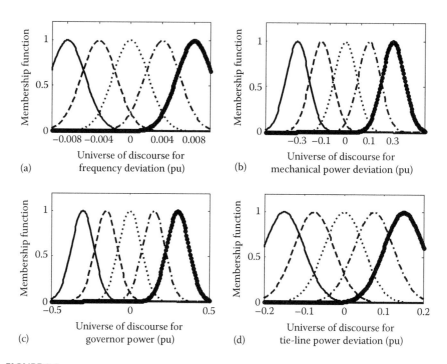

**FIGURE 9.2**
Membership functions for the state variables of area 1 and area 2 systems: (a) *x1*, (b) *x2*, (c) *x3* and (d) *x4*.

$$g_{b1}^{k}(X_1) = \frac{\prod_{r=1}^{4} \mu_{F_r}^{\ell_r}(x_{2r})}{\sum_{\ell_1=1}^{5}\sum_{\ell_2=1}^{5}\sum_{\ell_3=1}^{5}\sum_{\ell_4=1}^{5}\prod_{r=1}^{4}\mu_{F_r}^{\ell_r}(x_{2r})}, \quad k = 1,2...,5^4$$

$$g_{b2}^{k}(X_2) = \frac{\prod_{r=1}^{4} \mu_{F_r}^{\ell_r}(x_{1r})}{\sum_{\ell_1=1}^{5}\sum_{\ell_2=1}^{5}\sum_{\ell_3=1}^{5}\sum_{\ell_4=1}^{5}\prod_{r=1}^{4}\mu_{F_r}^{\ell_r}(x_{1r})}, \quad k = 1,2...,5^4$$

*Online adaptation*

1. The desired output is selected as $y_{d_1} = y_{d_2} = 0$ since the desired frequency deviation is zero.
2. Determine the tracking error vectors $\bar{e}_1 = \left[-c_{11}y_1 \ -c_{11}\dot{y}_1 \ -c_{11}\ddot{y}_1\right]^T$ and $\bar{e}_2 = \left[-c_{12}y_2 \ -c_{12}\dot{y}_2 \ -c_{12}\ddot{y}_2\right]^T$.
3. The components of the control signal are determined from (9.30), (9.31), and (9.46).
4. The controller parameters are updated online according to (9.44) and (9.45) with the design parameters $\gamma_{11} = \gamma_{12} = 5 \times 10^{-2}$ and $\gamma_{21} = \gamma_{22} = 5 \times 10^{-3}$.

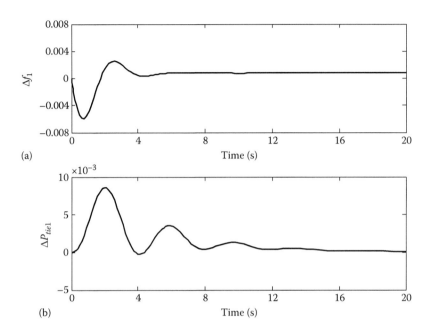

**FIGURE 9.3**
(a) Frequency and (b) tie-line power deviations for area 1.

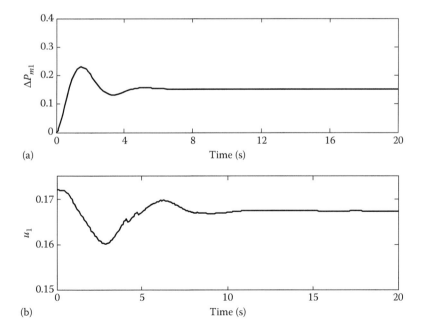

**FIGURE 9.4**
(a) Mechanical power deviation and (b) control signal for area 1.

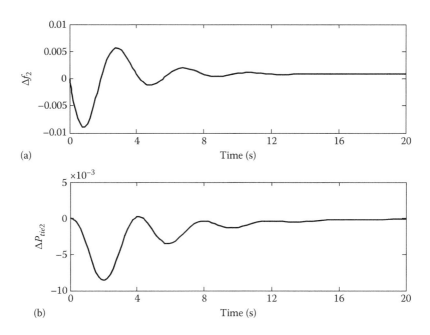

**FIGURE 9.5**
(a) Frequency and (b) tie-line power deviations for area 2.

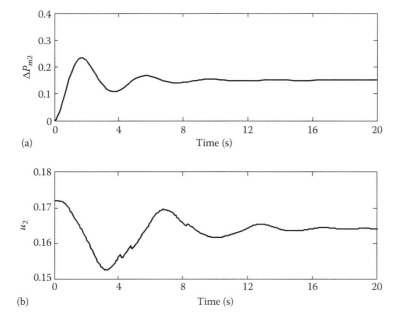

**FIGURE 9.6**
(a) Mechanical power deviation and (b) control signal for area 2.

A step load disturbances of $\Delta P_{d1} = 0.15$ and $\Delta P_{d2} = 0.15$ are used in the simulation of both areas fitted with the direct–indirect adaptive fuzzy controller. The simulation results are depicted in Figures 9.3 and 9.4 for area 1 and Figures 9.5 and 9.6 for area 2.

## References

1. C.-T. Chen, *Linear System Theory and Design*, Oxford University Press, Inc., New York, 2014.
2. R. L. Williams II and D. A. Lawrence *Linear State-space Control Systems*, John Wiley & Sons, Inc., Hoboken, NJ, 2007.

## Further Readings

1. L.-X. Wang, *A Course in Fuzzy Systems and Control*, Prentice-Hall International, Inc., Upper Saddle River, NJ, 1997.
2. H. A. Yousef, K. AL-Kharusi, and M. H. Albadi, Load frequency control of a multi-area power system: An adaptive fuzzy logic approach, *IEEE Trans. Power Syst.*, 29(4), 1822–1830, July 2014.

# 10

## Indirect Adaptive Fuzzy Load Frequency Control

### 10.1 Introduction

The human knowledge, in the form of IF-THEN rules, available to a control designer can be classified as plant knowledge and/or control experience. The former describes the plant behavior and the latter represents the required control actions in different operating conditions of the plant. In indirect adaptive fuzzy controller (IAFC), plant knowledge is used to construct a suitable fuzzy system. This fuzzy system is then used to model the unknown plant toward designing an adaptive fuzzy logic controller. The background of indirect adaptive fuzzy logic controller (IAFLC) was introduced in Chapter 8. Based on the IAFLC strategy, we will develop, in this chapter, a load frequency control (LFC) scheme for a multi-area power system.

### 10.2 Design of an IAFLC Load Frequency Control for a Multi-Area Power System

Suppose a power system consists of $N$ areas represented by

$$\left.\begin{aligned}\dot{x}_i &= A_i x_i + B_i u_i + \sum_{j=1, j \neq i}^{N} A_{ij} x_j + F_i \Delta P_{d_i} \\ y_i &= C_i x_i\end{aligned}\right\} \tag{10.1}$$

where the matrices $A_i$, $B_i$, $C_i$, $A_{ii}$, and $F_i$ are in the controller canonical form.

### 10.2.1 LFC Tracking Scheme for an Isolated Area

Each undisturbed isolated area is represented by

$$\dot{x}_i = \begin{bmatrix} 0 & 1 & 0 & 0 \\ 0 & 0 & 1 & 0 \\ 0 & 0 & 0 & 1 \\ -a_{0i} & -a_{1i} & -a_{2i} & -a_{3i} \end{bmatrix} \begin{bmatrix} \Delta f_i \\ \Delta P_{mi} \\ \Delta P_{gi} \\ \Delta P_{tiei} \end{bmatrix} + \begin{bmatrix} 0 \\ 0 \\ 0 \\ 1 \end{bmatrix} u_i \Bigg\}$$

$$y_i = \begin{bmatrix} 0 & c_{1i} & 0 & 0 \end{bmatrix} x_i \tag{10.2}$$

It can be shown that the relative degree of (10.2) is three, which means that the input $u_i$ appears in the third derivative of the output $y_i$, as indicated in the following equation:

$$\dddot{y}_i = C_i A_i^3 x_i + C_i A_i^2 B_i u_i \tag{10.3}$$

where $C_i A_i^2 B_i = c_{1i} \neq 0$. The $3 \times 1$ error vector $\bar{e}_i$ is defined as

$$\bar{e}_i = \begin{bmatrix} e_{1i} & e_{2i} & e_{3i} \end{bmatrix}^T = \begin{bmatrix} (y_{d_i} - y_i) & (\dot{y}_{d_i} - \dot{y}_i) & (\ddot{y}_{d_i} - \ddot{y}_i) \end{bmatrix}^T \tag{10.4}$$

where $y_{d_i}$ is the desired output signal. The error dynamic equation can be written as

$$\dot{\bar{e}}_i = \begin{bmatrix} \dot{e}_{1i} \\ \dot{e}_{2i} \\ \dot{e}_{3i} \end{bmatrix} = \begin{bmatrix} e_{2i} \\ e_{3i} \\ (\dddot{y}_{d_i} - \dddot{y}_i) \end{bmatrix} \tag{10.5}$$

Substituting (10.3) into (10.5), we obtain the following form of the error equation:

$$\dot{\bar{e}}_i = \begin{bmatrix} 0 & 1 & 0 \\ 0 & 0 & 1 \\ 0 & 0 & 0 \end{bmatrix} \begin{bmatrix} e_{1i} \\ e_{2i} \\ e_{3i} \end{bmatrix} + \begin{bmatrix} 0 \\ 0 \\ 1 \end{bmatrix} \left( \dddot{y}_{d_i} - C_i A_i^3 x_i - c_{1i} u_i \right) \tag{10.6}$$

The design of the tracking controller can be achieved by transforming (10.6) into Brunovsky form using the control law

$$u_i = \frac{1}{c_{1i}} \left( \dddot{y}_{d_i} - \Gamma_i(x_i) + v_i \right) \tag{10.7}$$

and hence the tracking error equation (10.6) becomes

$$\dot{\bar{e}}_i = \Lambda_i \bar{e}_i + B_i \begin{bmatrix} 0 \\ 0 \\ 1 \end{bmatrix} (-v_i) \tag{10.8}$$

where

$$\Lambda_i = \begin{bmatrix} 0 & 1 & 0 \\ 0 & 0 & 1 \\ 0 & 0 & 0 \end{bmatrix}$$

$B_i = \begin{bmatrix} 0 \\ 0 \\ 1 \end{bmatrix}$, $\Gamma_i(x_i) = C_i A_i^3 x_i$, and $v_i$ is a stabilizing control signal that locates

eigenvalues of the error dynamics in desired locations in the s-plane. This control signal is designed as a state feedback control in the form

$$v_i = \begin{bmatrix} k_{0i} & k_{1i} & k_{2i} \end{bmatrix} \begin{bmatrix} e_{1i} \\ e_{2i} \\ e_{3i} \end{bmatrix} = \overline{K}_i^T \bar{e}_i \tag{10.9}$$

Therefore, the closed-loop error equation becomes

$$\dot{\bar{e}}_i = \Lambda_{ci} \bar{e}_i \tag{10.10}$$

where $\Lambda_{ci} = \Lambda_i - B_i K_i^T$. The eigenvalues of the closed-loop system (10.10) are determined from the characteristic polynomial

$$\alpha_i(s) = s^3 + k_{2i}s^2 + k_{1i}s + k_{0i} \tag{10.11}$$

The control law (10.7) is not implementable when the parameters of (10.2) are unknown. Moreover, this control law does not take into consideration the interconnection and the disturbance terms.

Next, an indirect adaptive fuzzy logic LFC scheme will be designed for each area of an interconnected power system. The overall system is assumed to have unknown parameters and subject to constant load disturbances.

## 10.2.2 IAFLC Load Frequency Control Scheme

The interconnected system (10.1) can be written in the form

$$\left.\begin{array}{l} \dot{x}_i = A_i x_i + B_i u_i + M_i(X_i) + D_i \\ y_i = C_i x_i \end{array}\right\} \tag{10.12}$$

where

$$M_i(X_i) = \sum_{j=1, j \neq i}^{N} A_{ij} x_j = (A_{i1}x_1 + A_{i2}x_2 + \cdots A_{iN}x_N) = \begin{bmatrix} m_{11_i}(X_i) \\ m_{21_i}(X_i) \\ m_{31_i}(X_i) \\ m_{41_i}(X_i) \end{bmatrix}$$

$$D_i = F_i \Delta P_{d_i} = \begin{bmatrix} d_{11_i} \\ d_{21_i} \\ d_{31_i} \\ d_{41_i} \end{bmatrix}, \text{ and } X_i = \begin{bmatrix} x_1 & x_2 \cdots x_j \cdots x_N \end{bmatrix}^T_{j \neq i} \text{ is the composite state}$$

vector of the areas connected to the $i$th area. Similar to (9.27), the third derivative of the output is given by

$$\dddot{y}_i = \Gamma_i(x_i) + c_{1i}u_i + c_{1i}\chi_i(X_i) + c_{1i}d_{41_i} \tag{10.13}$$

where $\Gamma(x_i) = C_i A_i^3 x_i$ and $\chi_i(X_i) = (m_{41_i} + \dot{m}_{31_i} + \ddot{m}_{21_i})$, which accounts for the interconnection terms.

The IAFC is given by

$$u_{li} = \frac{1}{c_{1i}}\left(\dddot{y}_{d_i} - \hat{\Gamma}_i(x_i, \theta_i) - c_{1i}\hat{\chi}_i(X_i, \phi_i) + \overline{K}_i^T \overline{e}_i\right) \tag{10.14}$$

where $\hat{\Gamma}_i(x_i, \theta_i)$ and $\hat{\chi}_i(X_i, \phi_i)$ are fuzzy approximations of the functions $\Gamma_i(x_i)$ and $\chi_i(X_i)$ that are generated from the fuzzy system with a product inference engine, a singleton fuzzifier, and a center average defuzzifier. These fuzzy approximations can be written in terms of the fuzzy basis functions as

$$\hat{\Gamma}_i(x_i, \theta_i) = \theta_i^T f_{bi}(x_i) \tag{10.15}$$

$$\hat{\chi}_i\left(X_i,\phi_i\right)=\phi_i^T g_{bi}\left(X_i\right) \tag{10.16}$$

The parameter vectors $\theta_i$ and $\phi_i$ are adjustable according to updating laws to be derived later. The tracking error equation is determined as follows. From (10.14), $\ddot{y}_{di}$ is written as

$$\ddot{y}_{di}=c_{1i}u_{li}+\hat{\Gamma}_i\left(x_i,\theta_i\right)+c_{1i}\hat{\chi}_i\left(X_i,\phi_i\right)-\overline{K}_i^T\overline{e}_i \tag{10.17}$$

Subtracting (10.13) from (10.17), assuming $u_i=u_{li}$, we write the following error equation

$$\ddot{e}_{1i}=-\overline{K}_i^T\overline{e}_i+\left(\hat{\Gamma}_i\left(x_i,\theta_i\right)-\Gamma_i\left(x_i\right)\right)+c_{1i}\left(\hat{\chi}_i\left(X_i,\phi_i\right)-\chi_i\left(X_i\right)\right)-c_{1i}d_{41i} \tag{10.18}$$

Using the definition of the error vector $\overline{e}_i$ given in (10.5), this equation can be rewritten in the following state space form:

$$\dot{\overline{e}}_i=\Lambda_{ci}\overline{e}_i+\left[\left(\hat{\Gamma}_i\left(x_i,\theta_i\right)-\Gamma_i\left(x_i\right)\right)+c_{1i}\left(\hat{\chi}_i\left(X_i,\phi_i\right)-\chi_i\left(X_i\right)\right)-c_{1i}d_{41i}\right]B_i \tag{10.19}$$

Consider the optimal parameter vectors $\theta_i^*$ and $\phi_i^*$ defined as

$$\left.\begin{aligned}\theta_i^*&=\arg\min_{\theta_i}\left[\sup\left|w_{1i}\right|\right]\\\phi_i^*&=\arg\min_{\phi_i}\left[\sup\left|w_{2i}\right|\right]\end{aligned}\right\} \tag{10.20}$$

where $w_{1i}$ and $w_{2i}$ are the mismatch vectors defined by $w_{1i}=\hat{\Gamma}_i\left(x_i,\theta_i\right)-\Gamma\left(x_i\right)$ and $w_{2i}=\hat{\chi}_i\left(X_i,\phi_i\right)-\chi_i\left(X_i\right)$. The minimum mismatch vectors are given by

$$\underline{w}_{1i}=\hat{\Gamma}_i\left(x_i,\theta_i^*\right)-\Gamma_i\left(x_i\right) \tag{10.21}$$

$$\underline{w}_{2i}=\hat{\chi}_i\left(X_i,\phi_i^*\right)-\chi_i\left(X_i\right) \tag{10.22}$$

Adding and subtracting the terms $\hat{\Gamma}_i\left(x_i,\theta_i^*\right)$ and $c_{1i}\hat{\chi}_i\left(X_i,\phi_i^*\right)$ in (10.19) and then using (10.15), (10.16), (10.21), and (10.22) yields

$$\dot{\overline{e}}_i=\Lambda_{ci}\overline{e}_i+\left(\psi_i^T f_{bi}\left(x_i\right)+c_{1i}\phi_i^T g_{bi}\left(X_i\right)\right)B_i+R_iB_i \tag{10.23}$$

where $\psi_i = \theta_i - \theta_i^*$ and $\varphi_i = \phi_i - \phi_i^*$ are the parameter error vectors and $R_i = \left( \underline{w}_{1i} + c_{1i} \underline{w}_{2i} - c_{1i} d_{41i} \right)$ is a scalar quantity representing the minimum fuzzy approximation error and the disturbance term.

To determine the updating laws of the parameter error vectors, we choose Lyapunov function and find its time derivative along the tracking error trajectory as follows:

$$V_i = \frac{1}{2} \bar{e}_i^T P_i \bar{e}_i + \frac{1}{2\gamma_{1i}} \psi_i^T \psi_i + \frac{1}{2\gamma_{2i}} \varphi_i^T \varphi_i \tag{10.24}$$

$$\dot{V}_i = \frac{1}{2} \left( \dot{\bar{e}}_i^T P_i \bar{e}_i + \bar{e}_i^T P_i \dot{\bar{e}}_i \right) + \frac{1}{\gamma_{1i}} \psi_i^T \dot{\psi}_i + \frac{1}{\gamma_{2i}} \varphi_i^T \dot{\varphi}_i \tag{10.25}$$

Substituting (10.23) into (10.25), we get

$$\dot{V}_i = \frac{1}{2} \bar{e}_i^T \left( \Lambda_{ci}^T P_i + P \Lambda_{ci} \right) \bar{e}_i + \psi_i^T \left( \frac{1}{\gamma_{1i}} \dot{\psi}_i + \bar{e}_i^T P_i B_i f_{bi} \right)$$
$$+ \varphi_i^T \left( \frac{1}{\gamma_{2i}} \dot{\varphi}_i + \bar{e}_i^T P_i B_i g_{bi} \right) + R_i B_i^T P_i \bar{e}_i \tag{10.26}$$

The parameter updating laws are chosen as

$$\left. \begin{array}{l} \dot{\psi}_i = \dot{\theta}_i = -\gamma_{1i} \bar{e}_i^T P_i B_i f_{bi} \\ \dot{\varphi}_i = \dot{\phi}_i = -\gamma_{2i} \bar{e}_i^T P_i B_i g_{bi} \end{array} \right\} \tag{10.27}$$

Hence, (10.26) becomes

$$\dot{V}_i = -\frac{1}{2} \bar{e}_i^T \bar{Q}_i \bar{e}_i + R_i B_i^T P_i \bar{e}_i \tag{10.28}$$

where $\bar{Q}_i = -\left( \Lambda_{ci}^T P_i + P \Lambda_{ci} \right) > 0$.

To show that the tracking error $\bar{e}_i$ is globally ultimately bounded, we rewrite (10.28) in the form:

$$\dot{V}_i \leq -\frac{1}{2} \lambda_{\min} \left( \bar{Q}_i \right) \|\bar{e}_i\|^2 + \alpha_i \lambda_{\max} \left( P_i \right) \|\bar{e}_i\| \tag{10.29}$$

where $\|.\|$ stands for the Euclidean norm and $\lambda_{\min}(.)$ and $\lambda_{\max}(.)$ represent the minimum and maximum of the eigenvalues. The constant $\alpha_i$ is positive such that $\|R_i\| \le \alpha_i$. It can be shown that the ultimate bound of the tracking error is given by

$$b = \eta_i p_i \left[ \frac{\lambda_{\max}(P_i)}{\lambda_{\min}(Q_i)} \right] \tag{10.30}$$

where $\eta_i = \alpha_i/\delta_i$, $0 < \delta_i < 1$, and $p_i = \sqrt{\lambda_{\max}(P_i)/\lambda_{\min}(P_i)}$. The details to reach to (10.30) have been given in Equations 8.45 through 8.50.

### 10.2.3 IAFLC Load Frequency Control Scheme with Tracking Error Attenuation

In this section, an additional control signal is introduced to the control law (10.14) in order to reduce the tracking error in an $H_\infty$ sense. The new control law will be

$$u_i = u_{Ii} + u_{ai} \tag{10.31}$$

where $u_{ai}$ is the control signal responsible for attenuating the tracking error. This control part is designed as

$$u_{ai} = \frac{1}{c_{1i} r_i} B_i^T P_i \bar{e}_{ii} \tag{10.32}$$

where $P_i$ is a positive definite matrix satisfying the Riccati-like equation

$$Q_i = -\left[ \left( \Lambda_{ci}^T P_i + P_i \Lambda_{ci} \right) - P_i B_i \left( \frac{2}{r_i} - \frac{1}{\rho_i^2} \right) B_i^T P_i \right] \tag{10.33}$$

where $\rho_i$ is a desired attenuation level chosen such that $2\rho_i^2 \ge r_i > 0$.

### Theorem 10.1

For each interconnected and disturbed LFC area given in (10.12), the IAFC given by (10.31), (10.14), (10.15), and (10.16) along with the parameter updating laws (10.27) ensures the boundedness of the tracking error and the parameter error if the attenuation control term $u_{ai}$ is chosen as in

(10.32). Moreover, the proposed IAFC achieves the following $H_\infty$ tracking performance index:

$$\int_0^\infty \bar{e}_i^T Q_i \bar{e}_i dt \leq 2V(0) + \rho_i^2 \int_0^\infty R_i^2 dt \tag{10.34}$$

*Proof:*

In the presence of the control part $u_{ai}$, the closed-loop error equation (10.23) becomes

$$\dot{\bar{e}}_i = \Lambda_{ci} \bar{e}_i + \left( \psi_i^T f_{bi}(x_i) + c_{1i} \varphi_i^T g_{bi}(X_i) \right) B_i + R_i B_i - c_{1i} u_{ai} B_i \tag{10.35}$$

Along the trajectory (10.35) and using the updating laws (10.27), the time derivative of (10.24) is determined as

$$\dot{V}_i = \frac{1}{2} \bar{e}_i^T \left[ \Lambda_{ci}^T P_i + P_i \Lambda_{ci} \right] \bar{e}_i + R_i B_i^T P_i \bar{e}_{ii} - \frac{1}{r_i} \bar{e}_i^T P_i B_i B_i^T P_i \bar{e}_i \tag{10.36}$$

Adding and subtracting the term $\dfrac{1}{2\rho_i^2} e_i^T P_i B_i B_i^T P_i \bar{e}_i$ on the right-hand side of (10.36), we get

$$\dot{V}_i = \frac{1}{2} \bar{e}_i^T \left[ \left( \Lambda_{ci}^T P_i + P_i \Lambda_{ci} \right) - P_i B_i \left( \frac{2}{r_i} - \frac{1}{\rho_i^2} \right) B_i^T P_i \right] \bar{e}_i$$

$$+ \left[ R_i B_i^T P_i \bar{e}_i - \frac{1}{2\rho_i^2} \bar{e}_i^T P_i B_i B_i^T P_i \bar{e}_i \right] \tag{10.37}$$

The last two terms of (10.37) can be put in the form

$$\left[ R_i B_i^T P_i \bar{e}_i - \frac{1}{2\rho_i^2} \bar{e}_i^T P_i B_i B_i^T P_i \bar{e}_i \right] = -\frac{1}{2} \left( \frac{1}{\rho_i} B_i^T P_i \bar{e}_i - \rho_i R_i \right)^T$$

$$\times \left( \frac{1}{\rho_i} B_i^T P_i \bar{e}_i - \rho_i R_i \right) + \frac{1}{2} \rho_i^2 R_i^2 \tag{10.38}$$

Using (10.33) and (10.38) in (10.37), we obtain the following inequality:

$$\dot{V}_i \leq -\frac{1}{2} \bar{e}_i^T Q_i \bar{e}_i + \frac{1}{2} \rho_i^2 R_i^2 \tag{10.39}$$

Integrating both sides of (10.39) from $t=0$ to $t=\infty$ yields

$$V_i(\infty) - V_i(0) \le -\frac{1}{2}\int_0^\infty \bar{e}_i^T Q_i \bar{e}_i dt + \frac{\rho_i^2}{2}\int_0^\infty R_i^2 \, dt \qquad (10.40)$$

Since $V_i(t)$ is positive, (10.40) implies (10.34). This means that the $H_\infty$ tracking performance index is satisfied. When the initial conditions are zero, $V_i(0)=0$, then (10.34) can be written as

$$\|Q_i\|\|\bar{e}_i\|^2 \le \rho_i^2 \|R_i\|^2 \qquad (10.41)$$

where $\|Q_i\|$ is the induced 2-norm of the matrix $Q_i$, which can be determined in terms of the singular value [1] as

$$\|Q_i\| = \sigma_{\max}(Q_i) \qquad (10.42)$$

where $\sigma_{\max}(Q_i)$ is the maximum singular value of $Q_i$. Using (10.41) and (10.42), the tracking error is written in the form

$$\|\bar{e}_i\| \le \frac{\rho_i \alpha_i}{\sqrt{\sigma_{\max}(Q_i)}} \qquad (10.43)$$

where $\|R_i\| \le \alpha_i$. Equation 10.43 indicates that the tracking error of each area is ultimately bounded and the $H_\infty$ tracking performance index (10.34) is achieved. Note also that (10.40) can be written in the form

$$V_i\left(\bar{e}_i(\infty), \psi_i(\infty), \varphi_i(\infty), \infty\right) - V_i\left(\bar{e}_i(0), \psi_i(0), \varphi_i(0), 0\right) \le \frac{\rho_i^2}{2}\int_0^\infty R_i^2 \, dt < \infty \quad (10.44)$$

Equation 10.44 indicates that the tracking error $\bar{e}_i$ and the fuzzy approximation error vectors $\psi_i$ and $\varphi_i$ are bounded for $0 \le t < \infty$. This concludes the proof.

A block diagram of the proposed indirect adaptive fuzzy LFC is shown in Figure 10.1.

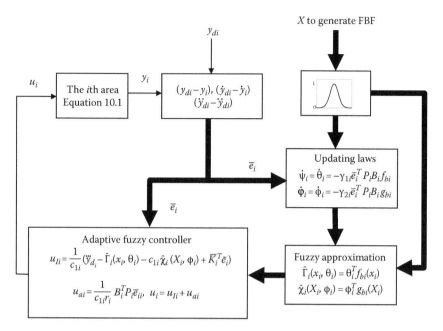

**FIGURE 10.1**
Block diagram of the proposed indirect adaptive fuzzy LFC.

### Example 10.1

Consider the two-area LFC system given in Example 9.1. The system matrices in the controller canonical form have been found in Example 9.1 as follows:

Area 1:

$$A_1 = \begin{bmatrix} 0 & 1 & 0 & 0 \\ 0 & 0 & 1 & 0 \\ 0 & 0 & 0 & 1 \\ -2 & -22 & -10.62 & -7.06 \end{bmatrix},$$

$$B_1 = \begin{bmatrix} 0 \\ 0 \\ 0 \\ 1 \end{bmatrix}, F_1 = \begin{bmatrix} 0 \\ -0.1 \\ 0.006 \\ 0.019 \end{bmatrix} \text{ and } C_1 = \begin{bmatrix} 0 & 1 & 0 & 0 \end{bmatrix}.$$

Area 2:

$$A_2 = \begin{bmatrix} 0 & 1 & 0 & 0 \\ 0 & 0 & 1 & 0 \\ 0 & 0 & 0 & 1 \\ -1.39 & -12.98 & -6.37 & -5.11 \end{bmatrix},$$

$$B_2 = \begin{bmatrix} 0 \\ 0 \\ 0 \\ 1 \end{bmatrix}, F_2 = \begin{bmatrix} 0 \\ -0.18 \\ 0.02 \\ 0.043 \end{bmatrix} \text{ and } C_2 = [0 \quad 0.69 \quad 0 \quad 0].$$

Interconnection matrices:

$$A_{12} = \begin{bmatrix} 0 & -0.69 & 0 & 0 \\ 0 & 0 & 0 & 0 \\ 0 & 0.14 & 0 & 0 \\ 0 & -0.008 & 0 & 0 \end{bmatrix}$$

and

$$A_{21} = \begin{bmatrix} 0 & -01.44 & 0 & 0 \\ 0 & 0 & 0 & 0 \\ 0 & 0.36 & 0 & 0 \\ 0 & -0.04 & 0 & 0 \end{bmatrix}.$$

The off-line and online steps to design an indirect adaptive fuzzy LFC for each area are summarized in the following:

*Off-line calculation*

1. The vectors $K_i^T = \begin{bmatrix} k_{0i} & k_{1i} & k_{2i} \end{bmatrix}$, $i = 1, 2$ are chosen as $K_1^T = K_2^T = \begin{bmatrix} 0.04 & 0.44 & 0.7 \end{bmatrix}$.
2. The positive constants $r_i$ and $\rho_i$ are selected as $r_1 = r_2 = 1$, $\rho_1 = \rho_2 = 1$ and the matrices $Q_1$ and $Q_2$ are assumed as $Q_1 = 20I$ and $Q_2 = 1.5I$.
3. Solution of the Riccati-like equation (10.33) yields the following positive definite matrices:

$$P_1 = \begin{bmatrix} 47.75 & 33.03 & 4.96 \\ 33.03 & 57.9 & 9.1 \\ 4.96 & 9.1 & 5.9 \end{bmatrix}$$

and

$$P_2 = \begin{bmatrix} 3.45 & 3.16 & 1.19 \\ 3.16 & 5.87 & 2.4 \\ 1.19 & 2.4 & 1.9 \end{bmatrix}.$$

4. Define the number of fuzzy sets for each state variable. Here, the same five Gaussian membership functions used in Example 9.2 for each state variable $x_i$, $i = 1, 2, \ldots, 8$ are

considered in this example. These membership functions are shown in Figure 9.2. Hence, the fuzzy basis functions $f_{bi}(x_i)$ and $g_{bi}(X_i)$ given in (10.15) and (10.16) will be similar to those used in Example 9.2.

*Online adaptation*

1. Since the desired frequency deviation is zero, the reference outputs are selected as $y_{d1} = y_{d2} = 0$.
2. Determine the tracking error vectors $\bar{e}_1 = \begin{bmatrix} -c_{11}y_1 & -c_{11}\dot{y}_1 & -c_{11}\ddot{y}_1 \end{bmatrix}^T$ and $\bar{e}_2 = \begin{bmatrix} -c_{12}y_2 & -c_{12}\dot{y}_2 & -c_{12}\ddot{y}_2 \end{bmatrix}^T$.
3. The components of the control signal are determined from (10.14) and (10.32).
4. The controller parameters are updated online according to (10.27) with design parameters $\gamma_{11} = \gamma_{12} = \gamma_{21} = \gamma_{22} = 1 \times 10^{-4}$.

The performance of the proposed indirect adaptive fuzzy logic LFC is simulated when step load disturbance of $\Delta P_{d1} = \Delta P_{d2} = 0.15$ occurs in both areas. The simulation results are shown in Figures 10.2 through 10.4.

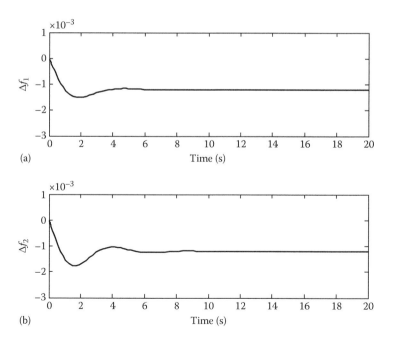

(a)

(b)

**FIGURE 10.2**
Frequency deviation (pu) of a two-area power system with IAFLC LFC.

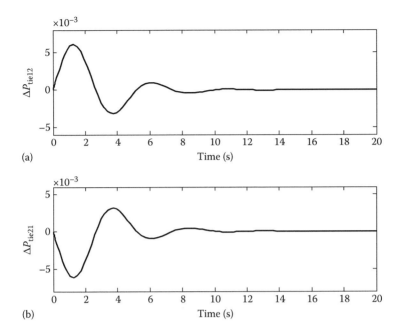

**FIGURE 10.3**
Tie-line power deviation (pu) of a two-area power system with IAFLC LFC.

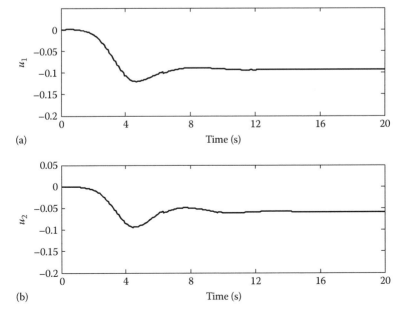

**FIGURE 10.4**
Control signals of a two-area power system with IAFLC LFC.

## References

1. F. Lin, *Robust Control Design: An Optimal Control Approach*, John Wiley & Sons Ltd., West Sussex, England, 2007.
2. H. Yousef, Adaptive fuzzy logic load frequency control of multi-area power system, *Electr. Power Energy Syst.*, 68, 384–395, 2015.

# *Appendix*

## MATLAB Scripts

```
% Example 2.1: Pole placement controller
An=[-D/(2*H)    1/(2*H)     0   ;
    0          -1/Tt     1/Tt ;
    -1/(R*Tg)  0        -1/Tg ];% system matrix
Bn=[0 ; 0; 1/Tg]; %input matrix
Fn=[-1/(2*H) 0 0 ]'; % disturbance matrix
Cn=[1 0.0 0.0]; % output matrix
DeltaPl = 0.2; %load disturbance
FnD =DeltaPl*Fn;
zeta =0.35, Ts = 4;sigma =4/Ts;wn=4/(Ts*zeta);wd =
wn*sqrt(1-zeta^2);% transient specifications
Pn=[-sigma+j*wd;-sigma-j*wd;-5*sigma]; % desired eigenvalues
Kn=place(An,Bn,Pn);% state feedback gain matrix
sys=ss((An-Bn*Kn),FnD,Cn,0);% closed-loop system with
disturbance
[yn,tn]=step(sys,8); % step response

%-------------------------------------------------------------------------

% Example 2.2: Pole placement with integral control action
A = [An zeros(3,1);Cn 0.0]; % Augmented system matrix
B= [Bn;0.0]; %Augmented input matrix
F=[Fn;0.0]; %Augmented disturbance matrix
C=[Cn, 0.0]; % Augmented output matrix
FD =DeltaPl*F;
P=[-sigma+j*wd;-sigma-j*wd;-5*sigma;-10] % desired eigenvalues
K=place(A,B,P);
K1 =[K(1,1) K(1,2) K(1,3)]; % state feedback gain matrix
KI = K(1,4); % integral feedback gain matrix
sys=ss((A-B*K),FD,C,0);% closed-loop augmented system with
disturbance
[y,t]=step(sys,8); % step response
plot(tn,yn,t,y)

%-------------------------------------------------------------------------
```

```
% Example 2.3 :Observer design
Po =[-6+j*2;-6-j*2;-50];% desired eigenvalues of error
dynamics
L= place(An',Cn',Po);% Observer gain matrix

%-----------------------------------------------------------------------

%Example 2.4 :Observer-based state feedback pole placement
plus integral %control
Aclo= [An    -Bn*K1            -Bn*KI;
       L*Cn   An-L*Cn-Bn*K1   -Bn*KI;
       Cn     zeros(1,3)        0 ]
eig(Aclo)
Fclo = [FnD;zeros(4,1)]
Cclo = [Cn,zeros(1,4)]
sysclo = ss(Aclo,Fclo,Cclo,0)

tx=0:0.01:5;
x0 = [0.1 0 0 .05 .05 0 0]';
u=0.2*ones(size(tx));
[yx,tx,x]=lsim(sysclo,u,tx,x0)

figure(1)
plot(tx,x(:,1),tx,x(:,4))% frequency deviation actual and
estimated
figure(2)
plot(tx,x(:,2),tx,x(:,5))%mechanical power deviation
figure(3)
plot(tx,x(:,3),tx,x(:,6))%governor power
```

# Index